Information as Receptive Relation

This book aims to revolutionize information research by introducing a receptive relation understanding of information, which systematically unveils its fundamental characteristics: created ex nihilo, emergence, reciprocity and shareability.

Through a thorough exploration of organismic and sensory receptivity, the book establishes a mechanistic foundation for understanding the nature of information. It navigates the origins of biological information and leads readers into a new era of information studies. Offering a fresh perspective on the nature of information, it delves into its physical, digital, and ideational encodings, as well as the ideational system built upon them. The book sheds light on critical issues such as quantum manifestation of information and the fundamental laws governing the relationship between information and matter/energy. It also dispels common misconceptions about information and its role in the evolution of information civilization.

The book provides valuable insights into understanding artificial general intelligence and the mysteries of consciousness and life. It will be of interest to researchers and students of information philosophy, information science, and artificial intelligence.

Tianen Wang is a Professor of Philosophy at Shanghai University with a Ph.D. from Wuhan University. His current research interests concern an integrated philosophical and scientific investigation of information, big data, and artificial intelligence based on systemic research on theories of description and stipulation.

Xi Wang is a lecturer at the School of Computer Engineering and Science, Shanghai University. She received her Ph.D. degree from Hosei University, Japan. Her research interests include software engineering and machine learning.

Information as Receptive Relation

Tianen Wang and Xi Wang

Routledge
Taylor & Francis Group

LONDON AND NEW YORK

First published 2024
by Routledge
4 Park Square, Milton Park, Abingdon, Oxon OX14 4RN

and by Routledge
605 Third Avenue, New York, NY 10158

Routledge is an imprint of the Taylor & Francis Group, an informa business

British Library Cataloguing in Publication Data
A catalogue record for this book is available from the British Library

ISBN: 978-1-032-77680-4 (hbk)
ISBN: 978-1-032-77805-1 (pbk)
ISBN: 978-1-003-48485-1 (ebk)

DOI: 10.4324/9781003484851

Typeset in Times New Roman
by Deanta Global Publishing Services, Chennai, India

Contents

Preface   *vi*

1   Introduction   1

2   A Receptive Relation Understanding of Information   25

3   The Quantum Manifestation of Information   51

4   The Veiling and Unveiling of Information   73

5   Information and Information Encoding   98

6   The Principle of Identity for Information-Matter in Operation   117

7   Basic Characteristics of Information   131

8   Information and the Development of Information Civilization   149

References   *171*
Index   *175*

Preface

This is an exhilarating era, where the fundamental nature of human existence and development faces a historic turning point. As fundamentally an information agent at the top level, the present moment for humanity is intricately linked with the unprecedented opportunities and developments in information technology. Grasping the rapidly evolving future increasingly involves a deepened understanding of information.

Understanding information involves a genuine foundation of shift from entitative paradigms to relational paradigms. For carbon-based humans, the information paradigm likely resides at the deepest level, reflecting its undeniable difficulty in transformation and paramount importance.

The challenge in the transformation of the information paradigm stems from its foundational premise concerning the different ways of human existence. The importance of the information paradigm is rooted in unraveling the mysteries of consciousness and life, thus implicating breakthroughs in the core mechanisms of human-like general intelligence. With the generalization of artificial intelligence, in-depth research on information becomes increasingly urgent. The deepening of information research, entangled with the mysteries of consciousness and life, not only addresses fundamental questions in philosophy and science but also involves the integration of philosophy and science, fundamentally impacting the further advancement of human self-awareness.

Whether dealing with artificial general intelligence or the mysteries of life and consciousness, typically involves an integrated understanding of philosophy and science, and information is not only an integration of philosophy and science itself but also the fundamental basis for the integration of all disciplines. This book aims to undertake information research on the integration of philosophy and science at the receptive relation hierarchy, laying an informational foundation for unlocking the core mechanisms of artificial general intelligence and the mystery of consciousness.

The fundamental paradigm shift that the receptive relation understanding of information involves is so foundational that neither science nor philosophy can independently achieve it. This shift surpasses the transformations introduced by the theory of relativity and quantum mechanics within the physics field, delving into a realm beyond matter and energy. Information research inherently calls for the integration of science and philosophy.

The receptive relation understanding of information, at a profound level, involves the comprehension of quantum physics, where quantum phenomena themselves wonderfully manifest information as receptive relation. In fact, all observational effects possess the same informational nature. In the macroscopic domain, due to the matching scale of human senses and objects, the receptive relations that serve as observational effects — information — can conveniently be attributed to the properties of objects. It is through this lens that information not only touches upon the fundamental concepts of all sciences but also enriches the understanding of human practices and cognitive activities.

As the most fundamental paradigm shift, the receptive relation understanding of information, in principle, also essentially involves the continual placement of all significant philosophical schools and their fundamental concepts. It is in the receptive relation understanding of information that we can witness the further unfolding of the relationship, epistemic and existential, between humans and the world, envisioning the broad prospects of human self-awareness.

This is a research field that demands thorough discussion, eagerly anticipating the exciting mutual inspiration it promises.

Tianen Wang
Shanghai
November 24, 2023

1 Introduction

In the present era, the age of information civilization is upon us, yet our understanding of the concept of information seems to be in a state that doesn't quite align with this new reality. Understanding the concept of information, which forms the foundation of the information civilization, is faced with a historic task.

Long before humanity even appeared on Earth, information was already present on this planet, yet our exploration of information has only spanned around one century. In daily life, we humans constantly deal with information, it is ubiquitous and surrounds us at every turn, but our understanding of information itself remains shrouded in mystery. The scientific community has conducted extensive research on information, while philosophical reflections on information have also been underway for quite some time. Today, the study of information has produced nearly 200 different definitions, but it appears that instead of achieving greater clarity on the concept of information, the fog surrounding it has only grown thicker. This peculiar situation in understanding information doesn't mean that the path is narrowing; quite the opposite, it suggests a unique cumulative effect in information research, hinting at the emergence of breakthroughs in information understanding and foreshadowing a moment of the dawn of research on information.

1.1 The Era of Information Research Integrating Science and Philosophy

Research on information has been explored comprehensively by both science and philosophy. However, it now faces a fundamental paradigm shift that neither science nor philosophy can independently accomplish. This shift goes deeper than the transformations brought about by the theory of relativity and quantum mechanics. The two pillars of modern physics brought about a paradigm shift in the field of physics, while information theory delves into a realm that is neither matter nor energy. Information research inherently calls for the integration of science and philosophy. With the development of information technology, particularly in the realms of big data and artificial intelligence (AI), what is now needed and possible at present is precisely the research on information in the integration of science and philosophy.

The scientific and philosophical research on information seems to indicate that even if we have not fully figured out what information is throughout the

DOI: 10.4324/9781003484851-1

understanding process of information, it does not seem to hinder the practical use of information-based applications in almost all fields, resulting in a strong contrast between the importance of information and its tranquil understanding. However, as disciplines move from differentiation to intersection and from synthesis to integration, the state of academic separatism in information understanding is increasingly inadequate to address integrated problems presented by both practical and theoretical concerns, and it can no longer meet the needs of contemporary development. Without a foundation of integrated scientific and philosophical understanding of information, a comprehensive grasp of information remains elusive. It will not be possible to grasp the common direction and trend of breakthrough emergence brewing in existing research on information, and understanding about information will not be completely in place; without a more accurate understanding of information, enigmatic phenomena such as quantum effects that exist like "ghosts" will be difficult to understand and grasp more deeply; the general development of AI can only stop at large language models (LLM) represented by ChatGPT, satisfied with the generality brought by its unique extension ability of artificial intelligence driven by big data, it will never be able to reach artificial general intelligence (AGI) with truly understanding ability. This also means that we cannot conduct deeper research into the mysteries of consciousness and life, as well as almost all related questions. This, in turn, means that we cannot have a more profound understanding of ourselves.

The scientific and philosophical understandings of information are often expressed as scientific information theory and philosophical information theory respectively. While this might appear somewhat biased from an integrated perspective, the information theories from both domains create significant tension and have been advancing relatively along the development of information understanding, just ushering in the emergence of a more accurate understanding and indicating a moment of "coalescence."

Currently, many efforts have been made to develop a philosophical understanding of information and its integration with its scientific counterpart, but the most widely accepted definition generally remains the scientific understanding of information. Luciano Floridi suggests, "We know that information ought to be quantifiable (at least in terms of partial ordering), additive, storable and transmittable. But apart from this, we still do not seem to have a much clearer idea about its specific nature" (Floridi, 2004). Quantifiability is a typical feature of scientific theories, and this statement makes it clear that the concept of information is relatively well-defined within certain scientific fields. Due to Claude E. Shannon's historical position as the "father of information theory" in information science, "the only academically recognized theory of information was Claude E. Shannon's 'A Mathematical Theory of Communication' established in 1948" (Yixin, 2013, p.71). However, Shannon's information theory is not universally accepted across all fields, mainly because it primarily pertains to the science of communication, even though its theoretical foundations suggest that the concept of information extends far beyond communication science. It's just that communication science deals primarily with signals, not information itself, Shannon's information theory, which was generated and mainly used in communication science, is understood as communication theory.

Conducting researches involving the evolution of information, Manfred Eigen, a Nobel laureate in chemistry, has long believed that, "Information theory as we understand it today is more a communication theory. It deals with problems of processing information rather than of 'generating' information" (Eigen, 1971). In the current context and development trends, advancing the research on information further by integrating science and philosophy is not only a higher-level imperative but also a historical mission to deepen the understanding of information in the age of the information civilization. As big data and artificial intelligence are developing at a fantastic speed, deeper-level research on information is not only very necessary but also comes at a perfect time. Building upon the existing results in scientific and philosophical research on information, creating higher-level understandings of information, and deepening the understanding of "higher level understanding" itself should be an urgent task for current research on information based on the integration of philosophy and science.

The integration of philosophy and science for the understanding of information does not aim to establish a unified definition of information as a final goal; instead, it seeks to provide higher-level concepts and theoretical holistic guan zhao (观照) for complex information research. The concept of "guan zhao" is unique to Chinese philosophy, and there is no equivalent term in English. Its meaning is to illuminate the parts with the whole, similar but not exactly equivalent to "holistic perspective" or "in the light of the whole." The current state of scientific and philosophical research on information has already provided rich intellectual resources for further deepening our understanding of information, creating significant intellectual tension and mechanisms that could constitute a historic opportunity.

Scientific understanding of information mainly aims at practical application within scientific domains. In scientific understanding of information, the application scope of information is usually very clear. For instance, in communication science, information is relative to noise; while in philosophical understanding of information, noise can also be considered information. Philosophical understanding aims to provide a holistic guan zhao at a level as high as possible. Although philosophical understanding of information lacks empirical certainty as in science, it provides a broader vision. Let's still take communication science as an example, in that field, information can be measured by eliminating uncertainty; while in philosophical understanding of information, uncertainty is an important characteristic of information. A more profound understanding of information represents a breakthrough grounded in the integration of science and philosophy, forming a bidirectional cycle of understanding.

Research on information by integrating science and philosophy ultimately responds to the needs of practice, especially the development of scientific practice. The development of information technology has revealed increasingly acute challenges in information understanding, some of which have become evident as dilemmas in information research. One of the most typical and representative dilemmas is the "information conservation" problem. On the one hand, in natural sciences, such as physics and biology, information is understood in terms of matter and energy. This understanding is not overly complicated, as information not being physical would imply that scholars in quantum physics would not be physicists, and those in information biology would not be biologists. On the other hand, not only do philosophers argue that information cannot be conserved, but also in daily life, the non-conservation of information is an unquestionable fact.

While the deepening of the understanding of consciousness and its mechanisms may not seem as urgent due to the already established state of human consciousness, the growing importance and bottleneck in the development of artificial intelligence make the understanding of consciousness and its mechanisms increasingly vital. As information civilization advances into the era of intelligent civilization, the understanding of information becomes an essential and pressing issue. The interconnection between information research and artificial general intelligence, the mysteries of consciousness and life, and their underlying relationships not only underscores the integrated nature of scientific and philosophical research but also highlight the complexity of understanding.

The requirement to understand information in the development of the integration of science and philosophy undoubtedly adds complexity to our explorations on one hand, but on the other, it provides a higher vantage point that allows us to better understand information as we navigate the developments of this new era. Some researchers who are well-versed in these mysteries have already realized,

> It is important here to understand that there is no reason to feel intimidated by the sheer complexity of these various dimensions of information. After all, it is a characteristic feature of our world that it is largely unknown to the human mind. Actually, throughout the course of history, the journey of mankind appears to be a perpetual quest in which we try to acquire a more complete understanding about this world, attempt to keep this acquired understanding safe, and make an effort to pass this understanding on from one generation to another.
>
> (Schuster, 2017, p. 5)

The advancement of information technology, especially in the realms of big data and artificial intelligence, is gradually maturing the conditions for the integration of scientific and philosophical understanding of information.

By utilizing big data to achieve a deeper holistic understanding and leveraging artificial intelligence to unfold information at higher levels, a foundation for a superior understanding of information has been laid. As people continuously deepen their understanding of information, a systematic exploration of the integrated levels of science and philosophy regarding information can be achieved. Due to the holistic nature of information involved, this exploration possesses deeper significance while simultaneously implying an extraordinary and challenging paradigm shift. Nonetheless, the issues posed by the era always signify the gradual maturation of corresponding era-specific conditions. The contemporary development of information technology is continually unfolding information.

With the development of information civilization, it is crucial to keep an eye on the progress of information understanding and continue to promote the development of information research. And in the understanding of information, what is currently most eye-catching is the understanding of basic concepts and properties of information, i.e., whether information is a mysterious existence with a veil or the emperor's new clothes, and the unfolding of information research based on paradigm shift implied by the current progress of information understanding.

1.2 The Nature of Information Concepts as Holistic Guan Zhao Series

The concept of information has been defined in nearly 200 different ways, making it a phenomenon-level problem. This phenomenon of understanding information itself carries significant epistemological implications. It allows us to understand the concept of information and its nature at a deeper level based on a holistic guan zhao of the concept series to which information understanding points.

1.2.1 Phenomenon of Information Concepts

As long as a general understanding of the relevant developments is gained, one will notice that "Information and talk of information is everywhere nowadays. …In spite of all this, there is no accepted science of information" (Barwise & Seligman, 1997, p. xi). In fact, not only is there no universally recognized general information science, but even in the use of the concept of information, we are still in a state of half-knowledge.

> Our everyday language includes the word 'information' in a variety of different contexts. It seems that we all precisely know what information is. Moreover, the explosion in telecommunications and computer sciences endues the concept of information with a scientific prestige that makes supposedly unnecessary any further explanation. …However, the understanding of the meaning of the word 'information' is far from being so simple. The supposed agreement hides the fact that many different senses of the same word coexist. The flexibility of the concept of information makes it a diffuse notion with a wide but vague application. If we ask: 'What is information?' we will obtain as many different definitions as answers.
>
> (Lombardi, 2004)

In the recently published English version of *What Is Information*, translators Eric Hayot and Lea Pao mentioned in their preface that today, "information" is a concept with an extraordinarily broad reach, and is most intensely under intellectual discussion in the United States. (Janich, 2018, pp. ix, xvii) In America, discussions about "information" are most intense. Particularly noteworthy are some intriguing paradoxical developments at the most basic level in some foundational scientific fields regarding information research.

Due to the accelerated development of information technology, the status of information is becoming more and more prominent. "The concept of information as 'knowledge communicated' has a dominant position in modern human life …From an epistemological viewpoint, the information concept has also found extensive and important use in biology, physics, psychology, etc" (Tzafestas, 2018, p. 159). On one hand, "Information is even developing as an important element of concern within the natural sciences, particularly within physics and biology, causing some to refer to information as the new language of science." (Beynon-Davies, 2011, p. 176) On the other hand, even philosophers who focus on conceptual analysis often do not discuss in depth when using the concept of information.

> the concept of information is generally treated in one of two ways. It is either defined quite narrowly … such as in terms of the transmission of binary digits (bits), or it is taken for granted in the sense that it is used as an important term but its meaning is very poorly defined and understood. Hence, for instance, the recent natural science definition for information only considers certain aspects of what we shall consider information. …Therefore, although information is critical "stuff" it is extremely difficult 'stuff' to pin down; it is probably not even 'stuff' at all.
>
> (Beynon-Davies, 2011, p. 176)

Human research on information seems to have reached an unprecedented node. On one hand, information and its concepts are so fundamental that attempts are made to reshape basic disciplines with information. On the other hand, the understanding of information appears so enigmatic that it seems to lead to a quagmire, regardless of the approach.

The diversity of concepts about information has led people to an extreme view:

> there simply is no universally agreed definition of information. If this is the case, how then can we speak of the mass-production of information? …Perhaps, from a Platonic point of view, we may say that we are talking about instances of the idea of information, but that we do not really have a complete understanding about the idea of information itself, "yet."
>
> (Schuster, 2017, p. 10)

On one hand, there is a belief that there isn't a unified concept for information. On the other hand, the term "information" is being used uniformly across an increasingly wide range, and information is even being mass-produced. There must be some reasons worth further exploration.

The reason for the lack of a universally accepted definition of information is multifaceted. The individuality origin from the specific existence of information is one of its distinguishing features. Concerning the concrete existence of information, it cannot be easily generalized using a dichotomous approach, as we can with the existence of physical objects. Our understanding of the concept of information has not yet reached a sufficiently high holistic level, which is obviously an important reason. Furthermore, at a deeper level, this may be closely related to our understanding of the nature and function of concepts.

As Ludwig Wittgenstein stated, language games have no essence; objects without abstract essence certainly cannot be generalized by dichotomy. Language games inherently possess informational properties and are, in fact, existence of informational ways. And the so-called "no essence" is actually that the specific existence of objects is so diverse (of course it's relative to specific agents) that cannot be dichotomously categorized, and therefore, we cannot achieve the kind of abstract generalization that can only be obtained through dichotomous distinctions, thus not reaching this level of abstract "essence." Wittgenstein thus dealt with language problems with family resemblance concepts, and the further problems brought up by this are: what kind of philosophical – or more specifically – epistemological consequences does the difference in abstraction level between dichotomy abstraction and family resemblance concepts have? Are degrees of abstraction not only the difference between dichotomy concepts and family resemblance concepts but also a series? These questions not only cannot be solved in traditional metaphysics but also lead to predicaments. Only by starting from such a level can we avoid predicaments and naturally reach a reasonable conclusion: the starting point and ultimate goal of abstract construction are to provide holistic guan zhao for the understanding of concrete things rather than abstract universality understanding itself.

The concept of information must – and certainly can – reach a definition at the highest possible level of universal utility (not universal acceptance, which involves two different philosophical views), because this is a tool with a holistic guan zhao function for understanding concrete existence of information under all specific situations, and tools always need to be and can always be produced and rationalized as much as possible. For research on information, the integration with science is the real mission that philosophy should undertake.

The fact that nearly 200 definitions were given for information not only indicates a lack of consensus on the understanding of information but also suggests the complexity and even uniqueness of understanding information itself. Some experts even argue that it is impossible to have a unified definition of information. "there simply is no single, universally acknowledged definition of information" (Schuster, 2017, p. 4). Even if there is a unified information theory, it is not necessarily helpful for information research. "It is quite possible that the general information theory will not bring any benefit to some practical problems, and they have to be tackled by independent engineering methods" (Stratonovich, 2020, p. 1). Due to the involvement of different disciplinary fields, it is indeed unlikely to achieve a traditionally unified definition of information in the scientific sense. However, this does not mean that a unified philosophical definition of information is impossible, or more precisely, the view that there cannot be a unified definition of information suggests that the philosophical concept of information is in urgent need of exploration. On a philosophical level, any concept should strive to achieve a higher level of understanding because philosophical understanding is not intended – nor is it possible – to reach the ultimate truth of abstract universality. Instead, it is meant to provide theoretical holistic guan zhao for information research in various fields, and higher-level concepts serve as the foundation for such guan zhao.

If a definition of information at a certain stage of development is not yet possible, it indicates that the depth of understanding has not been reached, and the necessary holistic level has not yet been attained which is often closely related to the process of cognitive development. Regarding this understanding, Dretske's perspective is intriguing: for philosophical purposes, something better is needed. It is not that philosophy is so much more precise, exact, or demanding than its scientific cousins. Quite the contrary, in most respects it operates with far fewer constraints on its theoretical flights. Nonetheless, the problems that define the study of philosophy are uniquely sensitive to conceptual issues. Getting straight about what one means is an essential first step. Words are the tools of philosophers, and if they are not sharp, they only disfigure the material. (Dretske, 1982, p.ix) In fact, for the integrative development of philosophy and science in the information civilization era, not only for philosophy, understanding of information requires theoretical holistic guan zhao from philosophy just like its requirement for local deep exploration can the systematic deepening of more holistic research about information be achieved.

1.2.2   Concept Series of Holistic Guan Zhao

The holistic nature of information is embodied in the increasing involvement of context factors at a higher level of information understanding. On one hand, due to its specificity, indeed, "information is a concept that appears in a wide range of contexts, each with its own specific motivations, observations, interpretations, definitions, methods, technologies, and challenges" (Schuster, 2017, p. 3). Consequently, people tend to believe that

> it is worthwhile to mention that it is surprising and very often overlooked that most investigations about information very quickly lead to a paradox. On the one hand, everybody seems to have a natural, intuitive, and immediate understanding about information. This understanding makes it relatively easy for any of us to appreciate, for instance, the important role that information holds as a driving-force behind our short-term actions and desires, as well as our more long term ambitions and goals. On the other hand, there is the simple fact that, as soon as we look at information more closely, it turns out that information is a rather elusive concept and that it is extremely difficult to pin down, exactly, what information "is".
>
> (Schuster, 2017, p. 4)

On the other hand, reaching a philosophical level of understanding of information can lead to a unified definition of information at a higher holistic level. Information research can be divided into several parts, one of which "deals with the philosophical side of information, and may involve discussions about the fundamental nature of information, how information may be generated, how it may become meaningful, or how information relates to moral issues such as responsibility or privacy" (Schuster, 2017, p. 4). The concept of information is indeed highly complex, leading Professor Zhong Yixin to propose the concept of an "information definition spectrum."

Given the multitude of information definitions and their inherent rationality, Zhong Yixin introduces an important concept, the "definition spectrum," into information research as the "information definition spectrum." The "definition spectrum" proposed by Zhong Yixin is a crucial reflection that develops to the new situation of defining the concept of information.

> According to the "strength of constraints" criteria based on definitions and the methods of systematic analysis, it is possible to establish an "order" for definitions within the complex web of information definitions. Based on this, if definitions with the weakest constraints (thus corresponding to the most general content and the broadest applicability) are placed at the forefront, and definitions with the strongest constraints (thus corresponding to the most specific content and the narrowest applicability) are placed at the end, an organized "information definition spectrum" can be obtained.
>
> (Yixin, 2013, p. 63)

In fact, the concept of the "information definition spectrum" is essentially a spectrum of definitions ranked according to the degree of abstraction under specific conditions. This concept holds significance not only for information definition but also for all definitions in general.

However, due to the ubiquity of information (often considered omnipresent in human activities) and its inherent complexity arising from various specific conditions, the concept of a "definition spectrum" is most typical for information definitions. Thus, the notion of a "definition spectrum" can indeed be established, but due to the practical nature of concept definitions in human usage, the majority within this spectrum may lack practical utility. Therefore, the concept of a "definition spectrum" is not as convenient as its representative counterparts, such as "philosophical definitions," "scientific definitions," and "everyday definitions." As understanding deepens, unfolding the "definition spectrum" as needed becomes not only reasonable but also necessary.

Understanding the concept of a "definition spectrum" is of significant importance, as it provides a crucial basis for shifting the understanding of concepts from reflecting the reality of objects to holistic *guan zhao* in the understanding of concrete things. The concept series is a holistic *guan zhao* understanding of abstract concepts which is entirely distinct from the pursuit of abstract concepts in the ultimate sense of universality.

In the case of specialized scientific research that can be divided into classes and divisions, the information concept in discipline integration at the holistic level only provides a holistic *guan zhao* for understanding concrete things in various specific discipline fields; while in information research with an integrated disciplinary nature, the same treatment leads to new issues. Since information encompasses applications in all domains, each application field must have sublevel information concepts that are more suitable for its specific conditions. The relationship between the top-level information concept and these specific information concepts in individual disciplines is akin to the relationship between universal philosophical concepts and special scientific concepts. This is a new form of concept that has emerged with the development of information research. This series of concepts, established based on specific conditions at different levels, can be referred to as a holistic guan zhao concept system.

1.2.3 Holistic Guan Zhao Mechanism of Information Concept Series

As a concept series, holistic guan zhao concepts are the result of abstract generalization of a single object in different concrete conditions that creates a collection of concepts with varying extensions. This type of conceptual form is distinct from conceptual systems, which consist of different concepts at different levels, and from family resemblance concepts, which describe the same object in clusters of concepts at the same level, focusing on different aspects of it respectively, while the holistic guan zhao concept series is composed of the same concepts at different levels. The holistic guan zhao concept implies concept series at different levels providing holistic guan zhao for understandings in the domain of each level. It is such a concept series that ensures the layer-upon-layer implementation of theoretical holistic guan zhao in different fields and at different levels. Oxford philosopher Luciano Floridi, who has made significant contributions to the field of information research, played an important role in this aspect.

Floridi's philosophical inquiries into information have raised profound questions that serve as crucial references for deepening our understanding of information and have significant implications for the initiation of information studies. His understanding of the current state of information involves delving into deeper philosophical layers. Concerning the question of "what is information?" Floridi put it this way:

> This is the hardest and most central problem in PI and this book could be read as a long answer to it. Information is still an elusive concept. This is a scandal not by itself, but because so much basic theoretical work relies on a clear analysis and explanation of information and of its cognate concepts.
>
> (Floridi, 2011, p. 30)

In this regard, the question of what information is resembles "what is the world?" However, what sets it apart significantly is that the question of what information is might be the first fundamental philosophical question that we can truly clarify. This highlights the necessity of the highest-level information concepts, which is closely related to the unified information theory problem.

Floridi raises further questions, such as "Is a grand unified theory of information possible?" On this issue, Floridi leans toward the view that it cannot be resolved in a reductionist sense (Floridi, 2011, p. 33), indicating more possibilities. He explains, "I personally side with Shannon and the non-reductionist" (Floridi, 2011, p. 33). The specificity of this question implies the rationalization of the way it is posed and the necessity of transformation. Reductionism not only implies an ontological premise but also implies a theoretical perspective connected to ontology. In fact, in terms of existentialism, with the development of cognition, as a prerequisite, the stipulation of "noumenon" orientating thought exposed its irrationality, and the view of theory based on ontological stipulation must undergo transformation: from treating theory as an ultimate goal of abstract universality grasped based on ontology to aiming at providing holistic guan zhao for specific understanding of concrete things. In this way of orientating thought, the question of "Is a grand unified theory of information possible?" can be understood in a new way. It is not a question of whether it is possible to make a truthful reflection based on objective grounds, but rather a question of whether it is meaningful and necessary to establish such a higher-level conceptual framework. Because any more accurate understanding of a specific thing must be premised on a higher-level holistic guan zhao, a unified information concept at a higher level is not only meaningful but also indispensable. It is just that the nature of the information concept and thus general concepts requires a new higher-level understanding.

The implementation of holistic guan zhao at various levels in the concept series reflects the hierarchical nature of theoretical holistic guan zhao. Within the concept series of holistic guan zhao, higher-level concepts, as a higher-level holistic guan zhao for lower-level concepts, and lower-level concepts offer insights that rationalize higher-level concepts. In conventional terms, the relationship between lower-level concepts and higher-level concepts is akin to subordination and superordination, while the relationship between the highest-level concept and its subordinate concepts is analogous to that between a constitution of a country and common law (as opposed to the constitution).

In different research domains at various levels, there must be corresponding-level conceptual holistic guan zhao tailored to specific conditions. This results in a series of hierarchical holistic guan zhao, constituting a hierarchy of theoretical holistic guan zhao in the development of interdisciplinary integration. Shannon's mathematical theory of information is a typical example; it is a concept primarily applicable in the field of communication science. Not only is it consistent with, but also provides insights into, understanding higher-level information concepts. Under the holistic guan zhao of higher-level information concepts, a deeper understanding can be achieved. Shannon's information theory not only establishes the foundational theory of information in communication science but also contributes to a deeper understanding of the information concept itself. Consequently, he believes that the concept of information "deserve[s] further study and permanent recognition" and elaborates on this by stating,

> The word "information" has been given many different meanings by various writers in the general field of information theory. It is likely that at least a number of these will prove sufficiently useful in certain applications to deserve further study and permanent recognition. It is hardly to be expected

that a single concept of information would satisfactorily account for the numerous possible applications of this general field.

(Shannon, 1993, p. 180)

This can be seen as a form of unconscious awareness of the holistic guan zhao concept.

At the level of integration of science and philosophy, theoretical holistic guan zhao should be a series that connects "Heaven" and "Earth," which means that two previously separate theories are docked: highly speculative theoretical systems and theories can be experimentally validated.

The concept of holistic guan zhao aligns with empirical research on theoretical models, providing a broad understanding of model interpretation. Models are closely related to holistic guan zhao.

Theories do not make contact with phenomena directly, but rather higher models are brought into contact with other, lower models (see chapter nine). These are themselves theoretical conceptualizations of empirical systems, which constitute an object being modeled as an object of scientific research.

(Floridi, 2011, p. 41)

The process of theoretical reasoning from top to bottom is based on models, which can be more precisely described as the process of holistic guan zhao.

Therefore, the holistic guan zhao process is ultimately based on models, and the model is the "shadowless lamp" and "hidden and spotlight" in the holistic guan zhao. On the other hand, the construction of theoretical models is a process of holistic development. However, the model possesses not only wholeness but also the nature and links directly connected to experience. It should occupy the forefront of understanding specific things the holistic guan zhao. It is a series of models between logic and experience that achieves the combination, even integration, of theory and practice. It is through concept series of the holistic guan zhao and theoretical models that we can unveil the layers of mist in the current use of information concepts at a deeper level.

1.3 A Mysterious Existence or the Emperor's New Clothes?

In the book *The Cult of Information*, Theodore Roszak, a history professor at the University of California, challenges the assumption that computers can provide multifaceted assistance. He questions,

Information has taken on the quality of that impalpable, invisible, but plaudit-winning silk from which the emperor's ethereal gown was supposedly spun. The word has received ambitious, global definitions that make it all good things to all people. Words that come to mean everything may finally mean nothing; yet their very emptiness may allow them to be filled with a mesmerizing glamour.

(Roszak, 1994, pp. xiii–xiv)

Roszak's viewpoint on the concept of information is intriguing. Rather than questioning the existence of information, he questions our understanding of it, reflecting objectively a brand new nature of the concept of "information."

The emergence of the concept of "information" is undoubtedly a significant milestone in the history of human terminology.

In philosophy, the concept of information has generated countless ambiguities and debates, as well as numerous misunderstandings, to the extent that it is difficult to find another word as important as "information" but with such an inaccurate understanding of its meaning.

(Feng, 2018, pp. 1–2)

Some even describe the concept of information as follows: "Information is a difficult concept to pin down but the arrows from the world to the eye and from the human to the hands capture something of what information is about." (Hardy, 2005, p. 67) In Roszak's perspective, we can find more intriguing viewpoints on information. He clearly pointed out that in "The Mathematical Theory of Communication," information theory established by Shannon is a science about information transmission. Roszak also mentioned a very important fact: Shannon separated "information" from usual facts. In his view, *A Mathematical Theory of Communication* established the discipline of information theory. Shannon's work is universally honored as one of the major intellectual achievements of the century. It is also the work most responsible for revolutionizing the way scientists and technicians have come to wield the word information in our time. In the past, the word has always denoted a sensible statement that conveyed a recognizable, verbal meaning which we would usually call "a fact." But now, Shannon gave the word a special technical definition that divorced it from its common-sense usage. In his theory, information is no longer connected with the semantic content of statements. Rather, information comes to be a purely quantitative measure of communicative exchanges, especially as those taking place through some mechanical channel which requires message to be encoded and then decoded, say, into electronic impulses. (Roszak, 1994, p. 11) In other words, the term "information" originally referred to "a sensible statement" that "conveyed a recognizable, verbal meaning," as a matter of fact, it means information encoding that people can directly understand, such as characters. However, characters cannot be transmitted directly, or rather, communication science cannot study the transmission of characters but must delve into more fundamental signals – namely, the electrical pulse encoding of information in communication science. This is why it is unrelated to the content being transmitted.

Roszak believes that from the beginning, Shannon was beset by the under-standable confusion that arose between his restricted use of "information" and the conventional meaning of the word. From his point of view, even gibberish might be "information" if somebody cared to transmit it. A message translated into a secret code would appear to be gibberish to anyone who barely knows the code, but it would be well worth sending to anyone who knows. The early information scientists easily fell into this way of thinking about messages and their transmissions; many of them had served as cryptographers during the war. Still, this was an odd and jarring way to employ the word, and Shannon had to admit as much. Roszak mentioned that when Shannon explained his work to famous scientists who questioned his strange definition of information, he responded,

> I think perhaps the word "information" is causing more trouble …than it is worth, except that it is difficult to find another word that is anywhere near right. It should be kept solidly in mind that [information] is only a measure of the difficulty in transmitting the sequences produced by some information source.
> (Roszak, 1994, p. 12)

There is no doubt that Fritz Mark Lup's description of information is most accu-rate: "infelicitous, misleading, and disserviceable." In his mind, it's the beginning of the term's history as "an all-purpose weasel-word."(Machlup, 1983, pp. 653, 658) This saying is not an exaggeration at all but very appropriate because, as a concept with fundamental paradigm implications, "information" actually means a major transition in human terminology development history. In this transition process, holistic guan zhao understanding of abstract conceptual nature occupies a key position.

The holistic guan zhao understanding nature of concepts not only provides a basis for understanding information concepts at higher levels but also highlights the necessity for a unified understanding of information concepts at higher levels, which puts forward historical tasks for deepening understanding of information concepts at a level where science and philosophy integrates.

The requirements for information research at the level of integration of science and philosophy are evidently in stark contrast to the current state of understanding and use of the concept of information. At a more foundational level, information is closely related to at least three iconic entities: DNA, which is closely associated with genetic information; data, and bits in digital technology; and the more com-plex forms of information in biological neural systems. It is these three domains that both constitute the richness of information and contribute to its extreme complexity.

The multiplicity of definitions for the concept of information is related both to the complexity of information itself and to the historical development of infor-mation understanding. Among the many definitions of the concept of informa-tion, the one given by Norbert Wiener holds a special place: "Information is information, not matter or energy" (Wiener, 1985, p. 132). The famous definition of information given by Wiener, the father of cybernetics, is regarded as a hall-mark of the information myth. His understanding of information is closely tied to cybernetics, and this famous statement is considered a concise representation of his cybernetic theories.

> The second important article of faith in the more general legacy of informa-tion theory, epitomized by the iconic claim that 'information is information, not matter or energy,' is Wiener's cybernetics, understood as a mathematical theory of control and feedback technology.
>
> (Janich, 2018, p.30)

This fundamental assertion about information serves as the starting point for almost all information research, both historically and logically. Since information is nei-ther matter nor energy, the question "What is information?" is evidently not only significant but also required to be addressed both empirically and logically.

Regardless of whether there is a universally accepted definition of information, there is a widespread consensus that our understanding of the concept of informa-tion needs further exploration. Information researchers have long recognized that there are diverse interpretations of the information concept.

> This obviously does not imply that an attempt to create a more general theory of information should not be pursued. On the contrary, it should. The existing theory is actually too restrictive. The first steps toward a more general theory are likely to be imprecise and loose, and they will probably irritate many people familiar with the neat and clear-cut nature of what they call a theory. But this is by no means a reason for dismissing the search.
>
> (Longo, 1975, p. iv)

The concept of information is indeed highly complex, and a historical examination of the development of information shows that

> over the past thirty years, information has unleashed human potential to a great extent, creating value that surpasses the sum of wealth accumulated over the previous five thousand years, but "information" remains a term that everyone is familiar with yet has a vague meaning.
>
> (Jun, 2013, p. v)

In fact, what's even more intriguing is that when discussing information, Shannon initially used "intelligence" instead of "information," which to some extent sug-gests a richer understanding of information.

Shannon used "intelligence" instead of "information" in the specific context of discussing what is transmitted in communication. "Off and on I have been work-ing on an analysis of some of the fundamental properties of general systems for the transmission of intelligence, including telephony, radio, television, telegraphy, etc." (Shannon, 1993, p. 455). The term "intelligence" has a long history and a very rich and complex set of semantics, seemingly more typical than "information" in conveying the complexity of this important concept.

In summary, there are myriad perspectives on information, ranging from considering it as the "emperor's new clothes" to viewing it as a mysterious presence yet to be unveiled. This phenomenon itself indicates that, with the development of information technology and the human information civilization driven by it, there is an urgent need to further deepen the systematic study of the concept of information.

The term "information" does indeed require clarification, but it is evidently not an easy task. In the history of scientific thought, concepts such as "motion" and "energy" have undergone complex characterizations, and the concept of "information" is not even in the same category as scientific concepts like "motion" and "energy." As a concept at the most foundational level, the redefinition of the information concept is both more challenging and more important, especially in the current context of the development of information technology and the information civilization. The state of research on the information concept clearly demonstrates the extreme importance of deepening the understanding of information in the integration of science and philosophy. The unfolding of information in contemporary development increasingly indicates that information is not the "emperor's new clothes" but a higher-level existence waiting for further unveiling. Contemporary development has created the conditions for a critical juncture in deepening the understanding and advancing the process of information understanding, paving the way for a winding path in information research.

1.4 The Use and Understanding of Information

As the information civilization continues to evolve, the situation regarding our understanding of information has become increasingly evident. As far back as the previous century, it was clear that In thinking about information, one tends to think of something objective and quantifiable—the electrical pulses surging down a copper wire, for example —and, at the same time, of something more abstract, of the news or message that these pulses carry—something not so clearly objective and quantifiable. For many purposes this is a useful ambiguity. It allows one to speak, for example, of information being picked up, processed, and passed along to the higher cognitive centers where it is used to control an organism's response to its surroundings. (Dretske, 1982, p.ix) More precisely, our understanding of information lags far behind its practical use. The use of the concept of information and the growing detachment from its understanding are increasingly apparent, as exemplified by Dretske. "It is, of course, fashionable to talk about information" (Dretske, 1982, p. viii), but "It is much easier to talk about information than it is to say what it is you are talking about" (Dretske, 1982, p. ix). The former involves the use of the concept of information, while the latter pertains to information understanding. This inevitably draws attention to the relationship between quantum mechanics and its theoretical interpretations. Due to its more direct involvement in paradigm shifts, the use and understanding of the information concept hold a more immediate significance.

Systematic research on information began in the field of communication science, and as a result, it has always been influenced by communication science in the understanding of the information concept and its utilization. The understanding process of information by humans is rather unique. As Scott J. Muller discovered,

> Information is a primal concept about which we have deep intuitions. It forms part of our interface to the world. Thus is seems somewhat odd that it is only in the last one hundred years or so that attempts have been made to create mathematically rigorous definitions for information.
>
> (Muller, 2007, p. 1)

This is because the study of information could only truly begin with the development of communication science. It was communication science that provided a basic framework for information research.

Regardless of how information is understood, discussions about information are based on three key elements: the sender (or source), the receiver, and the signal – the hallmark of information in communication science. In communication science, these three elements are particularly typical: first, the receiver, understood as the receiving end of information in communication science; second, the source or sender, where the term "source" is more original in meaning and "sender" originates from the sending of signals in the communication process. Therefore, the source is generally understood as the origin of information but is specifically the sending end of information in communication science. Third, the signal, which in communication science is understood as the physical quantity of representing a message.

Closely related to the transmission of signals, communication science also involves the unique concept of a channel. It is the best evidence for the fact that the information research framework in communication science cannot encompass many aspects of information content. The most typical difficulty lies in understanding the informational activities of plants. The requirements of using information concepts in communication science do not necessarily involve a cross-disciplinary understanding of the information concept. Thus, in communication science, on the one hand, we can see the historical advancement of information research, and on the other hand, we can see the dissociation between the use and understanding of information.

The dissociation between understanding and using the concept of information, particularly in the field of communication science, not only serves as a typical example but also underscores the significance of using information in relation to understanding it, "not because this theory (in its standard interpretation and application) tells us what information is. It does not. It does not even try." (Dretske, 1982, p. x) This is because it primarily operates at the level of utilizing information. Just like in the case of matter and energy, using information and comprehending it are often at variance, but their implications are more fully unfolded within the realm of information. Whether it is matter and energy or information, using them is much easier than understanding them. In practical terms, initially, it's all about utilization, and understanding comes along with usage.

At first, usage is a prerequisite for understanding, and as it progresses, understanding gradually becomes a prerequisite for further usage – this is a reciprocal developmental process where usage and understanding are interdependent. However, unlike matter and energy, due to self-involvement, human utilization and understanding of information will go through a more challenging process, while also possessing clearer conditions for scrutiny.

With the advancement of big data and the next generation of artificial intelligence, questions have become sharper, shedding light on the winding path of deepening information theory. Some more profound inquiries about information that have emerged from this have opened up valuable avenues of thought. It is through the digital collection, storage, and processing, particularly the information encoding highlighted by digitization and big data, that the potential for deepening the understanding of information, and even specific pathways, is implicit. This enables individuals to step out of their situation of own involvement as the most complicated receivers, allowing for a higher-level holistic examination of information.

1.5 The Winding Path of Information Research

Due to the unique connection between information and humans, on one hand, it's unlikely for us to fully grasp the understanding of information until information technology has advanced to the stage of big data and artificial intelligence. On the other hand, because of the uniqueness of information, human understanding of it has undergone a gradual and deepening development through its practical use.

Information research started within the framework of communication science, which has historical and logical origins and rationality. From this point, the understanding of information can expand toward both the biological information and the cutting-edge information phenomena in information technology, achieving a more in-place understanding of information. When researching on information, people first paid attention to the signal, then the source and the receiver, and finally gradually reached a more in-place understanding of information. Hence, there have been tendencies to understand information primarily through its signals, sources, and receivers. Due to their respective emphasis on different aspects, each approach has contributed to in-depth exploration, resulting in research in various directions regarding information. These studies are not only indispensable for the deepening of understanding but also directly emerge understanding of information with a paradigm shift nature based on these studies – receptive relation understanding of information.

Because different disciplinary fields have their own nature in understanding information, the scientific understanding of information tends to be rooted in specific disciplines rather than achieving a higher-level holistic understanding. In contrast, philosophical interpretations of information, based on the achievements of these disciplines, aim to reach the highest level of information understanding. This leads to three fundamental approaches to information understanding: signal-carrying understanding, source-preexisting understanding, and receiver-assigning understanding. These three basic approaches to information understanding serve as "pillars," constituting a "bridge" of the receptive relation understanding of information, and realizing a paradigm shift from the material paradigm to the information paradigm.

As paradigm shifts, information is understood as a process of receptive relation or relational processes. Chapter 2, based on the exploration of approaches to understanding information, provides a systematic discussion of the concept of information as receptive relation. Starting from the framework of information research in communication science, it builds upon the mature concept of information by characterizing it in terms of sensory receptivity. Expanding from sensory receptivity to organismic receptivity not only encompasses information activities in plants and more primitive organisms but also allows for the understanding of other human information activities beyond those based on sensory receptivity.

When we trace the history of human cognition through the lens of understanding information as receptive relation, we come across a fascinating fact in the process of development. In fact, long before quantum physics, information was already being typified as receptive relation. Observations in the quantum realm are receptive interactions, and quantum phenomena themselves are quintessential examples of information as receptive relation. On one hand, this understanding of information as receptive relation is clearly presented in quantum mechanics; on the other hand, it highlights the informational nature of quantum phenomena. Chapter 3 delves into the presentation of information through quantum phenomena and subsequently discusses the understanding of quantum physics and classical physics in terms of information.

What's even more intriguing is that despite the clear presentation of information as receptive relation in the realm of quantum physics, we missed this rare historical opportunity for information understanding for over a century. The further development of information understanding has now become possible, which is surely closely linked to the development of information technology but not solely depending on it. In fact, even today, the receptive relation understanding of information itself still brings serious problems for comprehension; it's destined to encounter insurmountable obstacles. A natural question arises: why receptive relation? This question has two main roots: it involves a fundamental paradigm shift and the unique human situation in understanding information.

In the development of human cognition, the receptive relation understanding of information involves new problems never encountered before. On one hand, understanding information involves the most fundamental paradigm shift so far. In the words of Wolfgang Hofkirchner, the paradigm shift of information is "A paradigm shift as profound as any in the history of humankind." (Hofkirchner, 2023, p.vii) Paradigm shifts have historically occurred within the realm of physics, from classical physics to the theory of relativity and quantum mechanics. However, the paradigm shift from the study of matter and energy to the study of information tran-scends the realm of matter and energy and extends into the domain of information, representing the deepest paradigm shift to date. On the other hand, humans find themselves in a unique situation in understanding information. Chapter 4 explores the unique situation for humans in information understanding and the resultant information obfuscation. It also discusses the development of information tech-nology, especially big data and artificial intelligence, and their role in revealing information.

Humans themselves are the most highly evolved receivers to date, and as receiv-ers, humans are deeply entangled in their own understanding of information. In the complex situation of deep entanglement as receivers, understanding informa-tion as receptive relation inevitably leads to the interpretation that everything that the receiver points to is information. Inertia in the material paradigm naturally causes information to be veiled by matter and energy. This even constitutes a mate-rial encoding ceiling in information understanding. It is this information encoding ceiling of information understanding that even veils the presentation of quantum phenomena as receptive relations. The development of information technology into big data and artificial intelligence increasingly reveals information in its full-ness. Big data, as a product of digital data development, is the evolving presence of informational existence collected through machine sensors. The generalization development of artificial intelligence points toward the level of information as receptive relation.

Marked by ChatGPT, large language models clearly present two phases of arti-ficial intelligence development: human knowledge-driven AI and big data-driven AI. The former is exemplified by expert systems and is limited to the integration of human knowledge due to the constraints of machine representation of human knowledge. The latter showcases unprecedented capabilities of generalization while still lacking the ability of human-like true understanding, hinting at a third-level development of artificial intelligence: information-driven AI. This illustrates the research path of AI developing from human knowledge to big data and further to receptive relation-based information-driven AI. Only at the level of information as receptive relation can AI achieve general and human-like intelligence, including perceptual abilities.

The revelation brought about by the development of big data and artificial intel-ligence highlights that the missing historical opportunity regarding the quantum phenomena manifestation of information is due, in part, to the veil of information understanding by the material paradigm and the adhesion between information and information encoding. The veil of information understanding by the material para-digm is closely related to the limitations in understanding information encoding brought about by the material paradigm understanding of information.

Without a sufficient understanding of information, it's impossible to grasp the intricacies of information encoding, and the relationship between information and its encoding accordingly, remains unclear. Based on the receptive relation under-standing of information, there is a theoretical premise for deepening our under-standing of information encoding. Chapter 5 focuses on the study of information encoding and its relationship with information. Information encoding is not merely the conversion between codes but the materialization and ideationalization of information as receptive relation. Accordingly, there are two fundamental types of information encoding: material encoding and ideational encoding. As symbolic coding in ideational encoding, digital coding has a special status. Since material encoding of information is the research object in all scientific fields relating to information, its research is more concentrated; ideational encoding of information is almost a completely new research area due to the tendency toward physical understanding of information, except for a special way in symbolic coding known as digital coding, which lays rich and important content about the evolution of information between material encoding and ideational encoding of information.

Through research on material encoding of information, not only can sources be understood more systematically but also evolution of information can be understood at a deeper level. As a concept relative to receiver, "source" implies that information sources encompass both primary and secondary sources. Primary information sources refer to all naturally concrete existences and their corresponding non-existences. The development of secondary information sources is intertwined with the development of receivers, constituting a bidirectional cycle mechanism. This gradually evolves into the concept of an "agent" that integrates information sources and receivers. With the development of agents and their information processing capabilities, secondary information sources emerge which are distinct from primary ones. Secondary information sources are generated through information processing by integrated agents, representing non-natural sources relative to primary sources. They encompass not only the products of processing primary information sources but also increasingly include potential sources (latent sources) created by agents.

As the level of information processing advances, secondary information sources become increasingly complex, allowing for continuous classification and upgrading as information production evolves. It is the development of secondary information sources and their classification, facilitated by the presence of big data as a substantial foundation that constitutes a crucial path in the transition of information understanding from a framework rooted in communication science toward a broader perspective.

Furthermore, research on ideational encoding of information not only derives a further conclusion that information is neither matter/energy nor concepts such as data and knowledge, but also can dock with the conceptual system established based on ideational encoding of information, and clarify the relationship between information as receptive relation and its encoding, as well as the ideational systems built upon them thereby deeply revealing the holistic guan zhao role of receptive relation in the establishment of ideational systems. It is through the ideational system that receivers have an important meaning-enduing mechanism in establishing deeper receptive relation.

Research on information encoding, especially material encoding of information, lays a foundation for deriving important and unique principles of information. Chapter 6 introduces the principle of identity for information-matter in operation, and by understanding and getting out of predicaments such as the conservation of information, it clarifies its important significance. As receptive relation, operations of information must ground on its material encoding, so they must comply with physical laws. It is, on one hand, the basis for the conclusion of "conservation of information," and, on the other, profound evidence for physical understanding of information. As law governing information, the principle of identity for information-matter in operation has obvious speciality compared with principles of matter and energy. The effective range of this principle is within the material operations of information; it only has operational significance, lacking ontological implications. The principle of identity for information-matter in operation serves as a deep-seated foundation for understanding the relationship between information and matter/energy. It not only has important significance for deepening understanding of matter and energy and even more research and resolution on integrated issues of science and philosophy but also lays a deeper foundation for systematically deepening the understanding of information; it involves systematic derivation of basic characteristics of information.

Adhesion between information and its encoding inevitably confuses the characteristics of information encoding with the characteristics of information itself. Confusing the characteristics of information itself with that of material encoding of information will inevitably lead to physical characteristic understanding about basic characteristics of information; therefore not only can basic characteristics of information hardly be systematically derived, but also the receptive relation nature of information can hardly be justified by the basic characteristics of information.

The receptive relation understanding of information and systematic research on the principle of identity for information-matter in operation laid a foundation for the systematic research on basic characteristics of information, which is totally different from that of matter/energy. Chapter 7 systematically derives the basic characteristics of information: created ex nihilo, emergence, reciprocity, and shareability. At the same time, it cuts off some adhesion with basic characteristics related to matter and energy.

As receptive relation, information is created ex nihilo, signifying a kind of creation out of nothing. This not only underscores the fundamental difference between informational resources and material resources but also highlights the higher-dimensional and hierarchical space that information development offers for human creativity. Creative activities are inherently human, and within the realm of matter and energy, human creativity is constrained to matter and energy transformation. However, within the realm of information, humans can engage in creative construction of information which means a kind of creative construction of holistic information product ex nihilo.

As receptive relation, information exhibits emergence at the level of generation mechanisms. Thus there are two basic types of emergence: emergence in the sense that functions or attributes cannot be reduced and emergence in that components cannot be reduced. The former is material emergence, and what it emerges is the new quality function that cannot be reduced. It's exactly the emergence of information that lays a deeper foundation for understanding the relationship between holism and reductionism.

As receptive relation, information develops from organismic receptivity to sensory receptivity, which signifies a transition from integration to the differentiation of receivers and sources, a process where reciprocity becomes more prominent. Reciprocity, as the interdependence, indivisibility, mutual influence, common action, and mutual stimulation among constituent parts of the same whole within a holistic process forms the foundation of the characteristics of information and, consequently, humankind characteristics as well as radical characteristics of anthropology. It is precisely the reciprocity of information imparting to information that needs a set of importance and qualities entirely distinct from those of material needs: the commonality of satisfaction and even germinating. Hence, the reciprocity of information holds profound significance for human characteristics and, consequently, for the development of humanity and human society.

The shareability of information is a basic characteristic that continuously unfolds from organismic receptivity to sensory receptivity, and it gradually becomes typical in the process of differentiation between the receiver and the source. As receptive relation, information can be generated by different receivers with the same source, so it has a true sense of shareability that matter and energy do not have. Precisely because of this, sharing in the sense of information is distinct from sharing in the sense of matter and energy. In Chinese, there are two relative concepts for representing the two meanings respectively: "gong xiang (共享)" and "fen xiang (分享)." In English, this differentiation between the characteristics of material and information can correspondingly be represented by "co-share" and "di-share." The word "di-share" implies that each party holds a portion, akin to the concept of "having a share in," such as apple and cake which can only be di-shared. On the other hand, "co-share" implies that all parties commonly own the whole, akin to "enjoying together," such as ideas and music which can be co-shared. Strictly speaking, matter and energy can only be di-shared, while information inherently possesses the characteristic of co-sharing. But due to the inherent association between information and matter-energy, the co-sharing nature of information can further develop matter and energy sharing.

As receptive relation, information possesses distinct fundamental characteristics on one hand and, on the other hand, these fundamental characteristics exhibit higher-level wholeness. Through the exploration of fundamental characteristics of information, we not only find confirmation of the understanding of information as receptive relation, but also recognize its profound and direct significance for the human information civilization and its development.

The systematic derivation and research on basic characteristics of information not only reflect the receptive relation understanding of information in turn, but also have great direct significance for the development of the human information civilization. The final chapter systematically explores the development of the human information civilization based on the receptive relation understanding of information. The evolution of human civilization ascends from the foundational level of material civilization to information civilization, it signifies a shift from primarily relying on material resources to primarily relying on information resources. Due to the basic characteristics of information, the development of the human information civilization entails a transition of society from di-sharing-based to co-sharing-based, with co-sharing being rooted in information. In the era of the information civilization, resource allocation increasingly shifts from ownership to usage, emphasizing that usage is the most effective form of ownership. The development of information civilization increasingly underscores the human need for information, and the commonality of satisfaction and germination of information needs lead to the growing superimposition of the reciprocity of information and anthropological reciprocity. Research into the information civilization and its development creates new conditions for the deepening of information investigation, forming a multi-level bidirectional cycle mechanism between the understanding of information and the development of the information civilization.

2 A Receptive Relation
Understanding of Information

2.1 Introduction

The advancement of contemporary information technology increasingly highlights a range of issues concerning the intersection of science and philosophy in information understanding. Among these, two prominent problems undergoing paradigmatic transformations are the dilemma of "information conservation" and the coding understanding of information. In the realm of natural science, "information conservation" is natural; otherwise, information processing would violate the law of energy conservation. This also shows the coding understanding ceiling of information. However, in philosophical research, information does not possess conservational properties; otherwise, even everyday phenomena would be inexplicable. Undertaking integrated research on the philosophy and science of information in depth is not only significantly important for resolving the dilemma of "information conservation" and the breaking through of the information coding understanding ceiling, but also for the development of artificial general intelligence (AGI).

We are currently living in an information civilization, yet our understanding of information concepts seems to be stuck in an inappropriate situation. While information science has experienced rapid advancement, our understanding of information has become a bottleneck, creating an intriguing paradox. It is quite intriguing that, despite an incomplete understanding of information, its applications in various fields proceed unimpeded. This results in an unusual balance between the importance of information and our level of understanding of it.

The growing disparity between the understanding and application of information has brought to prominence the hidden deep problem in the promotion of overall information research levels. The bottleneck of the core mechanism of artificial general intelligence, the striking contrast between big data application and its deeper understanding, and the unveiled mystery of the human brain reveal the importance and urgency of further understanding the concept of information at the fundamental level of human knowledge. The gradual unfolding of information by big data and AI highlights the necessity of upgrading information research to integrate philosophy and science.

For nearly a century, researchers have dedicated significant efforts to the study of information. As a fundamental concept, information is understood in various and occasionally conflicting ways. As a foundational concept involved in all fields

DOI: 10.4324/9781003484851-2

of recognition, information should be standardized differently according to the corresponding disciplines. However, conceptualizing information at a high level not only helps to deepen its understanding at a broader range, but also facilitates proper understanding of information for specific disciplines. Therefore, research on information needs to be advanced by integrating science and philosophy.

Significant efforts have been dedicated to understanding information from both philosophical and integrative perspectives. However, in general, the scientific understanding of information tends to dominate the definition in various fields. This is mainly due to the fact that current understanding is based on the mathematical description of information concepts, which aligns with the quantifiable nature. As a result, information concepts can be precisely described in most scientific disciplines. The reason why Claude Elwood Shannon's information theory has not been fully embraced across all disciplines is mainly because it is primarily a theory in terms of communication science. Only philosophical connotations can be derived from its theoretical framework. In fact, Manfred Eigen believed that "Information theory as we understand it today is more a communication theory. It deals with problems of processing information rather than of 'generating' information" (Eigen, 1971). Based on our accomplishments thus far, enhancing the integration of scientific and philosophical research on information is not only crucial for deepening our understanding of information, but also urgent for deepening our understanding of big data properties and the core mechanisms of the brain for the development of AI.

Understanding information from an integrative perspective does not necessitate a unified definition of information as the ultimate goal. Instead, it provides a broader framework for studying information concepts in the context of complex research. Scientific and philosophical research on information has already provided abundant resources for further deepening our understanding and has created promising ideologies that pave the way for historical progress. We are on the cusp of a breakthrough in information research.

The scientific understanding of information aims to serve practical purposes within the realm of science, while there is a higher requirement for its philosophical understanding. The former explicitly delineates its scope of application. For instance, in communication sciences, information is defined as the opposite of noise, whereas philosophically, noise can also be regarded as information. The latter aims to provide an overview at a preferably higher level; despite the lack of necessary certainty, it offers a broader horizon. Taking communication sciences as an illustration once again, information can be measured by the resolution of uncertainty. Yet, from a philosophical perspective, uncertainty itself can be treated as a key feature of information. Without the integration of science and philosophy, the fundamental paradigm shifts necessary for a proper understanding of information cannot be achieved.

Among the various definitions of information, a prominent one was proposed by Norbert Wiener, widely considered the founder of cybernetics. In his book, he stated, "Information is information, not matter or energy" (Wiener, 1985, p. 132). This statement, regarded as a cornerstone of cybernetics, has served as the basis for numerous information research projects due to its historical and logical validity. However, if information is neither matter nor energy, what exactly is it? This question holds significant importance, and addressing it through an interdisciplinary approach of science and philosophy is crucial.

Floridi has raised a series of profound questions regarding the integration of science and philosophy in information research. These questions serve as crucial references for deepening our understanding of information and are enlightening for researchers in the field. Floridi discusses the problem of using information to understand the status quo from a profound philosophical level (Floridi, 2011, pp. 30, 43). Regarding the fundamental issue of information, asking "What is information?" may seem similar to asking "What is the world?" However, unlike the latter, the former represents a fundamental question that can be more precisely addressed, thereby highlighting an important distinction between the two. Consequently, it underscores the significance of integrating science and philosophy in information research.

2.2 The Signal-Carrying Understanding of Information

Information is omnipresent for humans, much like air, which did not attract people's attention for a long time. The development of communication science highlights information, and information research formally began with the study of signals in communication science. Due to information research began with the focus on signals in communication science, the signal-carrying understanding of information appeared first. In communication science, the signal-carrying even signal understanding of information is taken for granted. If the information source is the generator of information, then the information channel is to information what a river is to a riverway, while the signal serves as a carrier of information.

The earliest understanding of information based on signal emerged in the field of communication. This brand new concept is developed in line with Shannon's work, wherein information is seen as the result of one uncertainty subtracted from another, which is exactly how communication science understands information.

Since the application of information started in communication science, it was also strictly defined first within this discipline. Here, the focus is on the process of signal transmission between the receiver and the source. Consequently, the definition of information in communication science centers on the signal and channel, regarding the signal being processed as information itself, leading to the signal-carrying understanding. This notion is evident in R.V.L. Hartley's paper "Information Transmission" (Hartey, 1928), where some argue that information is a "bit stream." If information is an objective and measurable phenomenon not "entangled" with humans, then as long as it is transmitted, even noises that are utterly meaningless to humans can be considered "information," according to Shannon.

The development of communication technologies has further strengthened the belief that information is carried by signals and can be treated as a signal or even seen as "signal" itself. As a result, many publications have equated "information" with "message" following the appearance of telecommunication technology, leading to the conclusion that "information is signal" (Yixin, 2013, p. 59). The signal-carrying understanding of information is further exemplified in Olimpia Lombardi's source-based information generation formula, where information is produced from a source, intuitively suggesting that information is defined as a signal. This is because communication refers to the transmission process of information, and the signal is the object transmitted during this process.

This provides a foundation for Wiener's first definition of information: "Information is a name for the content of what is exchanged with the outer world as we adjust to it, and make our adjustment felt upon it" (Wiener, 1989, p. 17). Wiener's further understanding of information is enlightening when acknowledging that it is neither matter nor energy. The phrase "what is exchanged with the external world" in the definition of information holds a deeper and more profound meaning for understanding than just from the perspective of the communication community. Simply viewing the signal as the object of transmission during communication fails to fully capture this deeper meaning, this is evident in Shannon's information theory.

A most precise understanding of information in communication science is the "elimination of uncertainty," derived from Shannon's information theory, from which we can trace not only features of communication science but also something that transcends. Shannon's information theory is undoubtedly a significant milestone in the further study of information theory. Whether it is directly developed based on it or further reflected upon through criticism, it has promoted the progress of information theory research.

An important indication of the signal-carrying understanding of information is the foundation it provides for quantifying information science. This allows communication information theories to be expressed mathematically (Schuster, 2017, p. 26). Therefore, the signal-carrying understanding is represented as a quantification understanding of information. The concept of information as a quantifiable entity originates from Shannon's *A Mathematical Theory of Communication*, where he discussed.

> Evidently the proper correction to apply to the amount of information transmitted is the amount of this information which is missing in the received signal, or alternatively the uncertainty when we have received a signal of what was actually sent.
>
> (Schuster, 2014, p. 26)

Accordingly, we conclude that "information is something for eliminating random uncertainty," which is actually an information understanding of message and a typical understanding of information from the perspective of communication science; since the message is, in fact, a series of signals. However, beyond the domain of communication science, information cannot solely eliminate uncertainty but can also introduce new uncertainties. At a philosophical level, certainty and uncertainty complement each other, making it impossible to conclude that information solely eliminates uncertainty. The change in specific quantity can be measured by mathematics in the elimination of uncertainty, involving the mathematical defini-tion of information in communication science and indeed belonging to scientific understanding of information. As a quantitative concept, this way of understanding is of great significance to be deeper mined and leads to a higher-level philosophical understanding of information.

Regarding Shannon's information concept as merely a quantitative concept will overlook the deeper connotations of Shannon's information theory. There may be different methods of quantifying objects, including within information theory, and measuring uncertainty in communication is one of the object quantification methods for information quantification in the communication field. However, it is inappropriate to regard one way of quantitative description as the qualita-tive description of information, even just for Shannon's information. Therefore, regarding Shannon's information concept as quantitative is a proper understand-ing of information in the light of communication science. However, for more general information concepts, Shannon's information theory encompasses much richer ideas that can be partly found in the research conducted by Fred Dretske on Shannon's information theory.

Dretske's semantic information theory was developed on the basis of Shannon's information theory, which extends beyond but remains confined to signal-carrying understandings of information due to the limitations of information's purely objec-tive position. This theory appears to be getting closer to quantitative measurement and its sentence structure, which fully demonstrates its scientific nature, has been extended from natural science to social science.

The quantitative research of information is mainly based on the signal-carrying understanding. The agreement on the existence of "information in terms of signal" indicates the treatment of signal as information. The signal-carrying understanding of information can even be clearly perceived through mathematical expressions.

Understanding information as signal typically regards information as something generated by the source and received by the receiver. "We apply the phrase quan-tum information to describe what we have received" (Richard, 2004). What the source generates, if any, is a signal rather than information. Otherwise, it would imply that at least some information is inherent in matter/energy. Considering quantum information as "the received information" in a quantum system is akin to regarding quantum information as a signal. As long as there is information embedded in matter/energy, it is impossible to differentiate information from mat-ter/energy. The mysterious nature of "quantum information" compared to ordi-nary information, stems from the mystery of signal within the quantum realm. In fact, quantum experiments are arranged in an integrative way, and the following questions remain unanswered: How can we understand signals during quantum measurement? What properties do those signals have? Does a signal in the classi-cal sense exist? Therefore, there is nothing fundamentally special about so-called "quantum information" which does not differ from common information. It is the uniqueness of signals in the quantum domain that makes "quantum signal" differ-ent in terms of signal-carrying understanding. The so-called mystery of "quantum information" stems from the mystery of quantum signals.

Not only is there physical evidence for the signal-carrying understanding of information, but there is also a connection to the feature of human intuition. Information is considered an abstract quantifier, similar to "water," which both characterizes and originates from the signal-carrying understanding of information. In the signal-carrying understanding of information, information is perceived like water, and "information flow" is logically utilized.

The abstract description property of quantum area, rather than quantum information, can be perceived between the non-matter property of quantum information and information as an abstract noun (Timpson, 2013, p. 4). However, this differentiation is based on interpreting Shannon's information concept as an abstract noun distinct from its ordinary usage. Regarding information as an abstract noun can blur the line between the abstract concept properties of information in information theory with its meaning, which indicates the information reflected by the actual process (such as communication). Contrasting with the signal-carrying understanding, regarding information as an abstract noun should demonstrate an effort to move beyond. It facilitates a deeper understanding of the signal-carrying understanding in the sense of language application.

Hereby we arrive at the concept of "information flow," where information resembles flowing water, and what we can obtain from this water-like relation between source and receiver must be the understanding of information as signal.

In this regard, the concept of information flow is based on the flow of signal, in other words, the flow of signal is regarded as the flow of information. Concerning the signal-carrying understanding of information, envision water flowing from its source to a reservoir. The source represents the origin of information, and the receiver stands for its destination – this embodies the signal-carrying understanding of information. When we consider the source of information as the source of water, and information as water, it becomes apparent why the saying goes, "The flow of information always points to the receiver." In fact, it's the signal under discussion. "Information flow" becomes a focus of study partly because signal is the object of research in communication science, where the flow of information is in fact the flow of signal.

With regard to the signal-carrying understanding of information, information exists at the source and flows from source to receiver, just like water flowing from its source to the sea. "Information flow" refers to, in fact, signal flow, just as "pheromone" essentially refers to the potential source. In the signal-carrying understanding of information, Dretske's perspective is representative, as evident in the title of his book *Knowledge and the Flow of Information*. The concept of "information flow" leads to the quantification of information, which, in turn, introduces the property of probability.

Another characteristic of the signal-carrying understanding of information is the probabilistic nature of signal processing, with probability being regarded as a property of information. In this vein, the understanding of signal as a way of understanding information is not yet mature. The understanding of information as signal is indeed simple and convenient, especially in the field of communication, but information and its carrier are confused. Under the condition that source and signal are all matter, information and matter can hardly be distinguished when confusing signal with information, despite its immaturity, the understanding of information as signal is of significance.

Firstly, the understanding of information based on signal represents a significant step forward the initiation of information understanding. Just as information research began with communication science, which holds an important position in information research, the understanding of information began with the signal-carrying understanding of information, which is crucial for information comprehension.

Secondly, the understanding of information based on signal is critical for the development of communication science. Inspiration about the signal-carrying understanding can be obtained from Floridi's description of the status quo of studies on information (Floridi, 2004, 2019). This, to some degree, explains why information is understood as signal. This aligns precisely with the understanding of information in communication science, where what communication engineering equipment and devices process is mainly signals with material properties.

Lastly, acknowledging the importance of signals themselves, the signal-carrying understanding of information holds great significance for understanding information. The transition from a signal in its basic form to a complex one establishes an important pathway for understanding information, moving from the signal-carrying understanding to a more proper understanding of information. In fact, a message is not information itself but rather composed of signals. To be more precise, information is neither signal nor message, and message is not information but a series of signals. Thus, it can be seen that understanding poses fundamental problems. As the first step in information understanding, the signal-carrying understanding of information inevitably encounters many problems. One of the most challenging issues is that considering information as something carried by signals encounters a paradoxical dilemma: information flow does not dissipate from its source like water flow. Hence, one can begin to sense the dilemma of the material paradigm in understanding information.

Since signal carries information from the source, the understanding of information in communication science developed and extended naturally to the source preexistence understanding. On the basis of signal-carrying understanding of information, it's natural to get to the further approach to information understanding: the source preexistence understanding of information.

2.3 The Source-Preexisting Understanding of Information

When information research transcends communication science, philosophical inquiries into information inevitably focus on the information source, leading to the emergence of the source-preexisting understanding of information. In fact, the source-preexisting understanding of information can be logically derived from the signal-carrying understanding of information. As a typical example, the idea of information being carried can be very easily generated in biological information activities. "Genetic information is carried from DNA to the ribosome by a messenger RNA molecule (mRNA)" (Avery, p. 47). However, due to the historical and logical development of science, philosophical inquiries into information originated from philosophical questions in information science, with the foundational starting point found in Wiener's work. Wiener's understanding of information serves as the basic starting point for further exploration. Building on Wiener's work, the question that naturally arises is: "What is information if it is neither matter nor energy?" This question naturally leads us to consider the "source" of information, resulting in a source preexisting understanding of information.

The source-preexisting understanding of information originates easily within the realm of science, particularly in physics and biology. In these fields, the focus is predominantly on physical objects, aiming to unveil the "information" inherent in these objects. The physical understanding of information represents one of the most complex and extreme forms of source-preexisting understanding. The most widely accepted view is that "Information is physical," initially proposed by German-American physicist Rolf Landauer. In relation to the source-preexisting understanding of information, Stonier's perspective is more systematic: "information has physical reality and constitutes an intrinsic property of the universe," "as real as are matter and energy" (Stonier, 1990, pp. 12, 107). Because of the closer relationship, the source-preexisting understanding of information has been more widely developed in the field of quantum mechanics. In 1991, Landauer published a seminal paper introducing the concept of "information is physical" within the context of quantum mechanics. The work of Landauer and subsequent efforts led to the establishment of the "quantum mechanical understanding of information theory" (Bub, 2005). Connecting quantum mechanics with information is of great significance, as it involves the core mechanism of linking information to physics, which can radically explain why information has gained increasing importance within the field. However, to fully comprehend this concept, a proper understanding of information is necessary.

Presently, the idea that "information is physical" remains a fundamental tenet in information theory. However, this concept poses challenges due to its relationship with matter/energy and the evolution of our understanding of information throughout history.

Firstly, considering information as an "intrinsic universal attribute of matter" can lead to a paradoxical contradiction. Unlike matter, information still needs to abide by the law of conservation of mass, even if we regard it just as a property of matter. This implies that equating information with matter would ultimately lead to the equivalence of information and matter.

Secondly, as the evidence shows, information is not necessarily a property of matter. To summarize information as a property of matter, ultimately it is still the material way of information understanding; therefore, ultimately, it is to summarize information as matter/energy.

Finally, a critical issue with the concept of preexisting sources as information is that matter/energy does not necessarily become the source of information, and the source itself may not be matter/energy – it could be non-existent. The existence of counterfactual communication in quantum communication serves as undeniable evidence.

Therefore, including information itself in the source not only leads to a better sources-preexisting understanding of information, but also results in a reversed genetic relationship between information and matter/energy.

Feng Xiao convincingly argued against understanding information as a property of matter. Considering pheromones as information itself is a typical method of source preexisting understanding and the best illustration of the challenges with this understanding, only because biology involves more complex information relationships, the problems concerning the source-preexisting understanding of information is not as obvious as in physics. By introducing the concept of the "Qubit ocean," Xiao-gang Wen concludes that "matter = information" (Bei & Xiao-Gang, 2019, p. 344). That is an extremely physical way of understanding information, information understanding, which took Qubit as quantum information that actually is information encoding.

The understanding of information has undergone, and continues to undergo, a process in which it becomes increasingly differentiated from matter and energy. Although certain aspects may overlap or be disordered due to different cognitive logics and historical reasons, it is overall a process of moving from matter/energy to its structure, organization, and form/pattern. Though identifying information as matter and proposing information physics essentially equate information with matter, claiming the objectivity of information holds great significance.

Organization encompasses a broader range than structure, and if we consider information as a degree of organization, the equation "matter = entity + energy + organization" can be modified to "matter = entity + energy + information" (Sokolov, 1991). In this context, it becomes challenging to imagine matter without information if we can achieve structure by integrating information into matter. If we cannot distinguish between matter and information, then what is the significance of saying "information is not matter"? If we reject the notion that information is not matter, how can we conceive of information that is not matter using the concept of "matter"? If information is not distinct from matter, then how can we explain the acquisition of information using the concept of matter? The problem of understanding the relationship between information and physics lies at the core of many issues concerning our understanding of information.

The challenge of source-preexisting understanding of information is more apparent in the field of quantum. "The claim that information is physical is baffling." On "Information is Physical," we noted that the slogan faced a difficult dilemma in the field of quantum information philosophy, whether it was supposed to advert information in the everyday sense or in the technical. In the field of quantum, the understanding of "information is physical" is severely challenged. Among the challenges, one of the incisive criticism came from Christopher Gordon Timpson, a professor in the Philosophy Department at the University of Oxford, which holds that, even in terms of purely linguistic or philosophical analyses, this predicate can hardly be established (Timpson, 2013, p. 72). When discussing the concept of "information is physical," considering the "identifier" as "physical" brings us closer to understanding information as physical encoding.

More scientific and modern efforts have been made toward categorizing information as either matter or energy. One approach, which is more scientific than representing information as a property of matter is understanding it as a field – where information is the unbalanced distribution of matter and energy in time and space. Thus, an information perspective claims that "information is the differences between entities," which connects back to the fundamental concept of information discussed by Wiener and Shannon.

In 1948, Shannon discussed the relationship between information and entropy, exploring the reduction of randomness and the increase in certainty of information. In 1950, Wiener built on Shannon's ideas by proposing that "information is negative entropy." The relationship between information and entropy is crucial and closely tied to the relationship between information and order. This correspondence aligns with the understanding of information in terms of order and is more relevant to matter/energy. However, whether we refer to "order" or "entropy," both concepts are related to the fundamental understanding of information as "difference." While considering information as something that eliminates uncertainty, negative entropy or difference alone does not provide a satisfactory philosophical definition, as information can also bring uncertainty, leading to entropy or zero difference. Nevertheless, each of these understandings possesses its own rationality as periodical achievements for information understanding, particularly Shannon's understanding.

In conclusion, source-preexisting understanding of information essentially claims that, as water flows from its source, information derives from an information source, making the source the generator of information. Information is a tangible entity that flows from the source to the signal, meaning it can only be understood as a physical entity in this context. Therefore, the source-preexisting understanding of information inevitably leads to an objective understanding of information, and the fact regarding information to be created by receivers naturally leads to repercussion.

2.4 The Receiver-Assigning Understanding of Information

The signal-carrying and the source-preexisting understanding of information, two points in a line, naturally point to the receiver-assigning understanding of information. In fact, the concept of "information flows from source to receiver" emphasizes the importance of the receiver-assigning understanding of information; it also highlights what is lacking in the understanding of information as source and signal. The crucial distinction between information and signal lies in the presence of meaning within information, whereas signal lacks inherent meaning. In communication science, the information theory can be approximately considered as a signal theory, but a deeper understanding of signals reveals the complexity of the issue. The involvement of receiver introduces another layer of complexity to the understanding of information. It is impossible to successfully transmit complex information from a complex source (e.g., a human) to a low-level receiver (such as a mouse) solely through the transmission of sophisticated signals. This undoubtedly demonstrates that information is not simply a "transmission" from source to receiver. This not only inspires the understanding of information as signal and source, but also establishes a foundation for the receiver-assigning understanding of information.

Although definitions of information based on the receiver can also be traced back to earlier times, they have received the least attention, which matches the development of human cognition.

If the receiver lacks understanding of or even fails to perceive the intended content of the received signals, specific information is not effectively received. Moreover, different recipients may derive varying or even contradictory meanings from the same received signals. This difference between information and signal indicates that information will not automatically flow into the receiver as signals. Thus, the receiver-assigning understanding suggests that information is produced by the assignment of the receiver.

The understanding of information based on the receiver comes into the picture as studies on the receiver and the receiver itself progress. The signal sent by the sender must be received by the receiver; otherwise, the signal is merely an electric pulse, light wave, sound wave, etc. In a desolate dessert, a desperate cry for help is not information if no one else could hear it. The series of signals, or "messages", sent into space by humans in an attempt to contact intelligent extraterrestrial life that might exist would be mere signals if not captured by alien agents. Only when intelligent alien life receives our signals does it compose information; in this way, there is a flow of signal from the source to the receiver. This illustrates the crucial role of the receiver in the information process; information understanding cannot be comprehended without considering the understanding of the receiver.

The reason why a receiver is a receiver lies not in its ability to store information like a water reservoir, nor is it merely a passive receiver. Instead, it actively perceives, and therefore, it is characterized by receptivity, the capacity to receive the source or signal. Therefore, the receiver is a key element in the composition of information and becomes increasingly important as it develops into a more advanced state. The development of the receiver, essentially, becomes the key point of information and can be regarded, to a large extent, or even radically, as the development of its receptivity which is closely related to the development from matter to information. Alongside the progression from matter/energy to information, the receiver in its development has gone through physical reactions, biological receptivity, and psychological receptivity. The emergence of computers has introduced physical receptivity, and information technologies, especially AI, play a significant role. With biological evolution, carbon-based life as the receiver, is becoming more and more intricate. Paradoxically, this complexity does not simplify the understanding of information for humans; instead, we appear to be further enveloped in the mystery of information. This complexity can be attributed to our involvement as intricate receivers in the cognition of information. It was only with the introduction of AI that the situation improved, as humans created machine receivers, leading to artificial informosomes (e.g., computers) and AI agents (e.g., intelligent robots). The development mechanism from matter to information can be revealed by unfolding the development of information technology. It is through the development of AI that we can look back at the evolutionary process of natural receptivity.

As a product of nature, the carbon-based receiver, or more precisely, carbon-based receptivity, has undergone two basic phases. The first phase involves the development of the receptivity of the most primitive organisms, represented by genes in DNA, the biological code of information. Gene being the original information matter agent of both source and receiver with no evolved receptors, is where matter/energy and information are in an integrative and meanwhile undifferentiated state. The absence of specialized receptive organs indicates that the receiver is undifferentiated from the source, mainly embodied in the physical state of decoding and encoding, completed via matter, i.e., through a biological process. The informational receptive relation process is undifferentiated from the material biological process, where matter/energy interactions coincide with information interactions, and the development of information overlaps with matter. This is a phase where matter/energy and information are intertwined: there are properties of both matter/energy and information, and there is matter/energy in the form of information as well as information in the form of matter/energy. The second phase of carbon-based receiver development involves a complex process from a biological receptivity base to a psychological receptivity base. Alongside the differentiation of the receiver and source, the carbon-based receiver has developed receptors and then evolved into receiver-source-integrated agents (informosomes) capable of processing and even generating information when the receiver itself became a source.

As a machine receiver based on physical receptivity, the silicon-based receiver was initially a silicon copy and extension of the carbon-based one. Up until now, the silicon-based receiver has developed from a camera to applications in today's next-generation AI. The development of machine receptivity provides important insights into the receiver-assigning understanding of information.

There is a close relationship between the receiver-assigning understanding and further research on the receiver. The complexity of the receiver itself indicates the complexity of the cognitive origin of the receiver-assigning understanding of information. One of the important aspects relates to a deeper understanding of the objectivity of information. To some extent, the receiver-assigning understanding is a reaction to the objective understanding of information. In contrast to understanding information in an objective way, the rebound from the obvious subjectivity of information prompts further exploration of the receiver, resulting in the receiver-assigning understanding of information. In a sense, it is the debate over whether information is objective or subjective that inspired the understanding of information based on the receiver.

The debate over whether information is objective or subjective concerns not only the understanding of information as source-preexisting and receiver-assignment but also involves the receptivity of the receiver, indicating the subject-object relations in the realm of information.

Due to the paradigm of knowledge developed through the study of matter, understanding information as signal-carrying and source-preexisting tends to objectify information to varying degrees and even leads to information objectivism in some cases. Since information is closely connected to matter, it should be treated as completely objective. In fact, the development of information is a process of differentiating objectivity and subjectivity, which closely relates to the differentiation of the receiver and the source. In the indifferent stage, information can be treated as entirely objective. However, once the source differentiates from the receiver, especially when the receiver has evolved into advanced agents or even multiple agents, the differentiation between subjectivity and objectivity emerges.

The receiver-assigning understanding of information, without leaving source or signal behind, rectifies a deviation in the understanding of information in an objective way; and the most typical point is an emphasis on the "human-dependent character" of information (Feng, 2018, p. 17). This explains why humans, as receivers, can delve deeper into information and why we seem to miss the whole picture when we are in it. Therefore the receiver-assigning understanding of information can be unfolded as a series understanding from high-level assignment to basic-level explanations. The emphasis on the "human-dependent character" of information represents a typical way of assignment understanding, while cognitive science serves as a typical representative of explanatory understanding.

The receiver-assigning understanding of information is prominent in cognitive science, where information is almost something that is aesthetically expressed by the receiver. As Dretske puts it, "Beauty is in the eye of the beholder, and information is in the head of the receiver" (Dretske, 1982, p. 55). Aesthetics does require special attention in understanding information. It can be seen that, to some extent, the explanatory understanding of information is receptive understanding; only the reception level has been upgraded by advanced human receivers under either explanatory or assigning understanding. This also explains the importance of the receiver-assigning understanding of information.

The receiver-assigning understanding of information has greatly contributed to research on the receiver, which is essentially the focal point of information research in the field. Undoubtedly, this understanding has its limitations. In contrast to the signal-carrying understanding of information, it is based on the negation of the source-preexisting understanding to varying extents and tends to ignore both source and signal. However, because of the special status of the receiver, the three understandings together at least objectively build a bridge for a more profound and accurate understanding of information.

In contrast to the source-preexisting understanding of information, which tends to be exclusive, the receiver-assigning understanding of information takes source into consideration to some extent and places excessive weight on the receiver primarily because of its focus on the human receiver. By "receiver," Feng Xiao also mainly refers to humans. Regarding information as an objective existence in the sense of matter would undoubtedly attribute information to matter, but understanding information solely as "human independent" would shelter the objectivity of information. A typical example would be animals that cannot make tools; and an extreme example would be the flush toilet, also known as "the beast of self" (Kelly, 1995, p. 109). To achieve a more precise understanding of information, the receiver plays a crucial role and the assigning understanding is an important step toward a better understanding of information

The over-reliance on the receiver in information understanding stems from both self-involvement within the human receiver maze and the fact that the receiver is the key to information research. As the highest-level receiver, humans are most critical to information research. This is natural to human kind. Moreover, research on the human receiver is also indispensable for a proper understanding of information. However, the key to information understanding lies not only in the highest-level receiver but also in getting the big picture of the receiver's development process. It is exploring the receiver in a broader and deeper way that greatly contributes to a profound understanding of information. Information does not exclusively belong to humans or even solely to living organisms. The receiver-assigning understanding of information not only holds the key to a more accurate understanding of information at a higher level, but also provides a higher-level overview of its receptive properties, thereby facilitating improvements in our understanding of information in various ways.

The understanding of information as source-preexisting focuses on the source and highlights the important correlation between information and matter/energy. The understanding of information as signal-carrying focuses on the signal and describes the process that starts when information is sent from the source; whereas the understanding of information as receiver-assigning reaches to the important discussion of how "information differs from matter/energy" and explores the fundamental properties of information that distinguish it from matter/energy. This discussion even extends to extreme scenarios without the source, representing an overall understanding based on its receptive understanding property. These three ways of understanding serve as individual piers in information understanding and steadily lead to its depth and build a bridge from matter to information, forming an important foundation for a more accurate understanding of information.

In conclusion, the receiver-assigning understanding of information is reasonable, but such a deep understanding comes at the cost of neglecting the objectivity of information. Indeed, the receiver can easily become the sole focus, as in "I think, therefore I exist" where the source is not involved, especially when regarded as advanced informosomes integrating both source and receiver. However, from the perspective of their relationship, the integration obviously indicates that there is more than just the receiver. Humans, as receivers, possess a complex form of integration of source and receiver, which is the fundamental reason for the receiver-assigning understanding of information to somewhat disregard the source and signal. Understanding information in the complex relationship between signal, source, and receiver is the key to getting rid of the solipsism tendency of the receiver-assigning understanding, where "the receiver is the only thing that matters" and achieving a more accurate understanding of information.

Further breakthroughs in information understanding on one hand must be based on the progressive research mentioned above and, on the other hand, it involves a basic paradigm shift. Although all three ultimately belong to the substantive paradigm of information comprehension, they have laid the foundation for the transformation of the information paradigm, which leads to relational understanding of information.

2.5 Understanding Information as Receptive Relation

If the signal-carrying or the source-preexisting understanding of information tends to regard information as something substantial and physical, then the receiver-assigning understanding of information tends to "create something out of nothing." Generally speaking, when we examine the understanding of information from the perspective of the signal, source, or receiver in isolation, they all result in a fragmented understanding of information relationships. However, due to the interconnection implied by the relationships among the signal, the source, and the receiver, many specific perspectives have enlightening insights into the relational understanding of information.

Considering signal flow as information flow is a mistake that is reasonable for use. It is a mistake not because of treating the abstract information as something concrete (though this attributional style holds significant logical importance), but due to understanding information, which is actually an effect, a process, or a relation, as merely signal-carrying.

2.5.1 Information Is Relational

In the study of information, some researchers focus on relation understanding, others address issues concerning relation, and some even clearly advocate understanding information as relation.

One way to understand information as relation can be seen in the intersection between "Relational Quantum Mechanics" (Rovelli, 1996) and "Information Quantum Mechanics" (Khrennikov, 2004). However, this understanding remains largely quantified which neither involves the qualitative aspects of understanding information as relation nor goes beyond the scope of physics. A typical example is the ordinary understanding. When we claim that information is relational, it is with respect to "the amount of information that can be extracted in practice from the data-set by a situated agent" (Boisot, 2007, pp. 26–27). In this case, the notion of "information is relational" apparently refers to the quantification of information rather than a qualitative understanding of information as relation.

With regard to understanding information as relation, qualitative understanding is a frequent guest to philosophical discussions. Floridi inspiringly put an emphasis on relation from three viewpoints which are transcendent to some extent: (1) Understanding information *as* entity i.e., ecological information, when information is regarded as a pattern of physical signal neither true nor false, and this goes beyond the signal-carrying understanding of information; (2) Understanding information *on* entity, when understanding of "semantic information" goes beyond the receiver-assigning understanding of information; (3) Understanding information *for* entity, when "hereditary information" goes beyond the source-preexisting understanding of information.

Evidence shows that Floridi, with his exploration on a philosophical level, already realized that information is relational (Floridi, 2011, p. 43). This is close to the relational understanding of information on the qualitative level

Among the studies on information, many have the tendency to understand information as relation to different degrees, though without explicitly stating it. For instance, Changlin Liu pointed out that information, as a dynamic process, involves the sending and receiving of information, reflecting a special relation between two entities, i.e., the relation between sending and receiving information (Changlin, 1985). In fact, the concept of "information" described here is no longer merely something transmitted that embodies "certain contents," instead, it can only be understood as a relation. This indicates a clear tendency to understand information as relation, although it has not yet reached the paradigm level; Changlin ultimately defines information as a property of entity: philosophically, it can be understood as a property of entity being reflected or a reflected property of an entity. This shift from a purely relational understanding back to an entity property understanding offers valuable insights. Understanding information as relation holds significant meaning in terms of a paradigm shift.

However, the scope of relation is so wide that it involves not only specific entities but also abstract concepts, and relation can be found or established between any entities or concepts; only understanding information as common relation is far from enough and will never involve relation paradigm shift. Since not all relations are information, a key question that needs to be further addressed arises: What kind of relation is information?

In understanding information as relation, significant progress has been made by considering the common connections and interactions among elements. This perspective is more concrete and precise than merely stating "information is a universal form of relation between things." Logically viewed, it tends to understand information as relation, thus directly regarding information as relation: "Information is relational" (Boisot, 2007, p. 27). Italian scholar Giuseppe Longo's statements in *Information Theory: New Trends and Open Problems* hold historical significance in the development of information theory, making huge progress in understanding information. He wrote: "Only differences (i.e. relations) matter" (Longo, 1975, p. iii), explicitly stating information's relational nature and considering relation as the starting point of information theory.

These works demonstrate that the receiver and source are the two basic factors that constitute information relation, complementing each other and being inseparable. If we further understand the relation from the development of the receiver's receptivity, we can conclude that the reason why the receiver is a receiver is because of its receptivity, and the reason why the source is a source is because of its perceptibility to a receiver. As the relation between receiver and source, the peculiarity of information as relation is closely related to receptivity.

2.5.2 Receptive Interaction

Due to its unique position in information understanding, as the most advanced and complex receiver, humankind is deeply immersed and inevitably resides in the information "Lushan."[1] In this context, a proper understanding of information is impossible without the development of information technology.

It is the information technology, particularly the development of big data and AI, that has led to groundbreaking revelations of information. From big data to AI, digital development has gradually unveiled the mystery of human receptivity. An electronic receptor is sufficient to demonstrate interactions in receptivity, distinct from pure matter in a straightforward way.

With regard to the interaction of matter/energy, electrons are very magical. At the molecular level, matter/energy moves in extremely small units with significant limitations on speed and space, akin to how "pheromones" function in ant societies. Electrons travel great distances in objects at speeds comparable to light. However, on the human scale, the interaction of micro electrons barely holds any meaning in terms of matter/energy on a macro scale. The lower the volt, the less meaning it holds on a macro scale and the more the electron current functions only as a signal.

For a signal to function, matter/energy must possess receptivity, that is, the capability to perceive the signal. The original form of this perceptual capability is the developed receptivity of an entity.

Receptivity can not only amplify interaction but also generate new-quality interactions, leading to novel styles of interactions between entities and new possibilities for entity development. An illustrative example is the development from strong-current interaction to weak-current interaction. As Kevin Kelly notes, "The amount of energy needed to send a signal is so astoundingly small that electricity had to be reimagined as something altogether different from power" (Kelly, 1995, p. 102). Their tricky difference is indeed related to the strength of entity interactions, but fundamentally lies in whether matter/energy has receptivity.

With receptivity, even an extremely small quantity can induce interaction, extending beyond purely matter/energy interactions and giving rise to two fundamentally different interactions: purely matter interaction and receptive interaction. About this, Fred Dretske provided a classical case of "knock door."

> When it is this pattern of knocks that causes you to believe that your friend has arrived, then (it is permissible to say that) the information that your friend has arrived causes you to believe he has arrived. The knocks might also frighten away a fly, cause the windows to rattle, and disturb the people upstairs. But what has these effects is not the information, because, presumably, the fly would have been frightened, the windows rattled, and the neighbors disturbed by any sequence of knocks (of roughly the same amplitude). Hence, the information is not the cause.
>
> (Dretske, 1983)

Extending the analogy of a knocking pattern to a secret signal aids in explaining receptive interaction. The receiver obtains important information from the specific knocking sound: the person who is coming is the designated receiver. What occurs between two designated receivers is entirely a receptive interaction. Similarly knocking on the door, as a factor, forms different interactions with different factors. The interaction with the neighbor is through the knocking sound, but the result is that the neighbor is disturbed, making receptive interaction primary, while physical interaction is secondary. Relative to the event of a fly being startled and flying away, the vibrations caused by knocking serve as the primary physical interaction while receptive interaction is secondary. Knocking the dust off a door through vibrations is purely a physical interaction.

The transition from purely entity interaction to receptive interaction is crucial for world development, marked by fundamental differences. Purely matter/energy interaction follows physical laws when entities interact. When matter receptivity is introduced, significant results can be achieved with minimal effort, leading to interactions unequal according to physical laws. If results beyond normal physical interaction can be achieved, the property of receptive interaction will be established. As a key factor in interaction, energy contributes more to receptive interaction than entity, performing better when serving as a signal.

The energy required to transmit an entity is proportional to its quantity, making the performance of an entity as a signal less effective. Energy consumption increases and performance worsens for entities of larger quantity. By adopting an energy unit smaller than the subatomic scale as a signal, lower energy consumption and better performance can be achieved. The smaller the energy unit is, the better the performance.

If the interaction between the observation instrument and an object is purely physical, then it is a physical interaction. If a specific part of the instrument involves a chemical mechanism, such as in taking pictures with a film camera, then it becomes an interaction with chemical properties. The transition from pure entity interaction to receptive interaction is established for interactions with chemical properties, such as those in a film camera. Receptive action is performed to some extent by both photoreception and data generation from an electronic receptive system. However, the distinctive property of receptive interaction is established when the cumulative effect of entity interaction becomes perceivable by human eyes. Thus, the presence of receptivity allows us to distinguish between entitative and informational interaction. Receptive interaction indicates its informational property, enabling us to clearly identify the difference between chemical photoreception and pure physical action.

The relationship between the interaction of matter and information will be an important research field. Informational interaction is always grounded in matter interaction, and there is no purely informational interaction without matter. Because informational interaction is always conducted through matter interaction, acting on matter through informational interaction holds particular significance.

When entitative interaction produces results beyond physical effects or when there is an effect that cannot be produced by matter and is thus inexplicable, informational interaction may be involved. The fundamental difference between informational and material interactions lies in the presence of receptivity in the former and its absence in the latter.

Receptivity arises from the capability to react, and reaction is the corresponding action induced by external influences, such as electromagnetic induction. Simple entities only have the capability to react, and the most basic level of reaction can be found in physical phenomena. In contrast, living organisms have the ability to perceive. For instance, enzymes and plants can perceive. Plants perceive sunlight, air, and water, enabling the development of biological receptivity controlled by DNA. The intrinsic connection between information and receptive interaction can be seen even in the most original DNA. Here, "the processor" is the receiver. If DNA alone does not suffice to manifest the receptivity of plants, consider *Drosera indica* or *Mimosa pudica*, whose behaviors are the most typical examples of plant receptivity and informational phenomena that we need to deal with, challenging the notion that only human brain activities are informational. Furthermore, an important feature of receptive interaction is evident in the receptive process. In interactions concerning receptivity, what is perceived as functions is more so as a signal than as matter/energy. An astonishing example is that some celestial bodies we see now have already annihilated several light years ago. A more common example is that the receptive interaction between source and receiver is mainly conducted through signal in the field of communication. In this process, the signal is not information but a key factor in establishing the receptive relation between the receiver and source, as signification is a crucial part of informational interaction.

Concerning receptive interaction between entities, the entity or factor that triggers the interaction already possesses a function similar to that of a signal, and it is only in receptive interactions that the perceived object or the trigger functions as a signal. For receptive factors, anything that can be perceived has the potential to be a source, and anything that can be perceived and sent by the receiver has the potential to be a signal.

A typical matter serving as a potential signal is the secretion molecule produced in the air by the ubiquitous pheromone in the insect world, indicating an important property of the informational process in chemical ways. There is a significant difference between interactions involving chemical properties and those involving physical properties. Chemical interaction is the key point in the development from material interaction with physical features to informational interaction with signal features, explaining the importance of chemical evolution in the evolutionary process. There is an important difference between acting on physical matter and chemical signals. It leads to the interactive relationship between physical interactions and signal interactions which is the original form of the relationship between material and informational interactions. The most significant difference lies in the fact that acting on matter requires the presence of matter, while acting on a signal can be achieved by an interaction without matter when the receiver is capable of reception. The reason is that the absence of a specific entity can also be treated as a signal to trigger certain reception, exemplifying the mechanism of counterfactual communication in quantum communication. In principle, any form of matter can act as a signal in a pure material interaction, but for rigid entities of larger scale, their interaction distance will become closer and therefore be dominated by physical features. Otherwise, their interaction distance will become greater and therefore be dominated by informational features.

The significant difference between interactions of physical and chemical properties stems from the fact that chemical interaction occurs at the atomic level. Chemical interaction serves as the key transition point in the development from material to informational interaction. Interaction at the atomic level is a typical example of the correlation between matter and information interaction where the distinction between these two kinds of interactions transforms from quality to quantity: interactions involving strong electricity are more likely of a material nature, while those involving weak electricity are more likely of a signal nature. Since signal interaction requires receptivity, interactions involving strong electricity are more likely of a material nature, and those involving weak electricity are more likely of a receptive nature.

Receptive relation does not require rigid contact between entities in receptive interaction. Pheromones can work even through interactions at the molecular level and over long distances. Another example is that the sense of sight and color obtained by light waves or photons, as well as the sense of hearing and smell obtained by the vibration of gaseous molecules, enables interaction between objects over long distances. The most typical form of material interaction is between solid objects, followed by interactions between liquid objects, whereas interactions involving gaseous objects are more likely of a signal nature. The movement of gaseous molecules is more flexible, and electrons at the subatomic level serve as the signal medium for a Turing machine. Electronic computers generate bits directly relating to matter via "on" and "off" states of electricity to form complex signals, thus leading to a typical informational interaction. Encoding signals with bits has only the digital forms of 0 and 1 as units; yet on the quantum level, where informational interaction is on the subatomic level, qubits can exist in a continuum of states between 0 and 1. Therefore, the factor acting as a signal is of critical significance within receptive interaction.

Receptive interaction is closely related to the matter/energy aspect, and it also enables the development of receptive relation. It is the receptive relation that serves as the focal point of Archimedes for the paradigm transition of information understanding.

2.5.3 Information Is Receptive Relation

Wiener's basic understanding of information as neither matter nor energy seems tautological without a clear definition, yet it effectively establishes what information is not. In the case of information with only matter on the same level, this is an extremely rich, profound, and inspired idea for understanding information and provides a wide space for further exploration. However, this prompts a crucial question: if information is not matter and cannot be separated from matter, what, then, is it? Numerous researchers have posed this question, building upon Wiener's foundational notion.

Compared to source-preexisting, understanding information as relation is a radical departure because only through this lens can information be distinguished from matter/energy. However, defining information merely as the relation of matter/energy falls short, as most such relations have nothing to do with information. Furthermore, information is not merely a relation, as the concept of relation is very broad and encompasses not only the relation of matter/energy but also conceptual relations. From this, we can hereby infer that information is a unique type of relation.

Expanding on Weiner's perspective, we can develop a receptive relation understanding of information based on the relation understanding of information and the receiver's receptivity. At its core, information is not matter or energy but a unique type of relation, the receptive relation based on matter/energy. Its typical state is the receptive relation between the receiver and source at its mature stage, representing a relation of a process relation or relation process.

"Recept" refers to perceiving stimuli, while receptivity denotes the characteristic of receiving and responding to stimuli. In the case of plants, receptivity to nutrients and sunlight is demonstrated through biochemical ways in which organismic receptivity to their environment. Animals and AI exhibit typical forms of sensory receptivity through eyes, ears, nose, tongue, body, and various sensors. The closing of a sensitive plant upon touch demonstrates the development of advanced sensory receptivity, while the Venus flytrap's hunting of insects can be seen as an early transitional form from organismic receptivity to sensory receptivity. To describe different relations, the meaning of "receptive" is significantly different from that of "perceptive", it is more primitive and occurs within the organism, in organismic process more precisely, while perception occurs through sensory processes, in sensor processes.

Receptive relation refers to the process of initial receptive effect or the reception of a source or signal from a sender by a receiver in the mature case, which can be observed in quantum phenomena or visual effects resulting from the interaction between a receiver and a source. Actually, when we view leaves and perceive the color green, this is an example of a receptive relation. AI sensors, like those used in autonomous cars, also establish receptive relation. The core mechanisms of AGI, life, and consciousness all deeply involve information as receptive relation. However, there remains a core mechanism issue between the establishment of receptive relation, AGI, and the mysteries of consciousness and life that needs to be addressed. Research in these areas must be based on the foundation of information as receptive relation.

Receptive relation is a process of receptive interactions. From the perspective of the mature state of information development, it is the receptive relation between a receiver and a source, with the receiver being the primary element in the receptive relation. Based on the receptive relation understanding, we suddenly realize that quantum mechanics is the most suitable field for understanding information. Considering that the observation involves the micro-scale, quantum mechanics provides unprecedented conditions for this kind of distinction. Human observation in the quantum field is a typical example of illustrating the relation between matter/energy interaction and information interaction, even the relation between matter/energy and information. As an observation effect or production, the "quantum phenomenon" itself is information as receptive relation. Even when looking back at macroscopic fields from the quantum mechanics perspective, we can even find daily examples inspiring our understanding of information. The green color of leaves in our eyes is itself information as receptive relation. The green color is neither an intrinsic property of leaves nor a feature of eyes, but an effect produced by the receptive interaction between the light wave reflected by leaves and our eyes. If the interaction stops, the green color as receptive relation will disappear simultaneously. In this sense, classical physics is the kind in which information (such as the green effect) can be ignored, or more precisely, taken as an inherent property of the leaf itself, and quantum physics is one in which information cannot be ignored. The next key question is how to understand information as receptive relation before the development stage of the separation of source and receiver. This involves the most critical field of receptive relation understanding of information.

For receptive relation understanding of information, biological information encoded in DNA plays a crucial role. It serves as both the final fortress to be conquered and the "promised land" for the paradigm transition in information. To understand information as receptive relation, the evolution process of information before the differentiation of the source and receiver becomes critically important.

Biological information, which predates the emergence of humans, is indeed critical to understanding information; for that, it poses both a significant challenge for interpretation and a crucial starting point for understanding.

With regard to studies of information, biological information holds a uniquely significant position. It involves fundamental questions that demand answers and is an integral part of unfolding information understanding itself. Marcello Barbieri's research on biological information is particularly noteworthy. His work started from the two major discoveries in molecular biology, not only narrating how information understanding develops in the field of biology, but also involving information paradigm transition.

Barbieri delved into the ontology of biological information based on two paradigms,

> today we have two conflicting paradigms in biology. One is the 'chemical paradigm,' the idea that 'life is chemistry,' or, more precisely, that 'life is an extremely complex form of chemistry.' The other is the 'information paradigm,' the view that chemistry is not enough, that 'life is chemistry plus information.' This implies that there is an ontological difference between information and chemistry, a difference which is often expressed by saying that information-based processes like heredity and natural selection simply do not exist in the world of chemistry.
>
> (Barbieri, 2016)

The understanding of these paradigms here is proper and critical. The "chemical paradigm" is the matter/energy paradigm closest to receptivity, from which the "information paradigm" differs a lot. The involvement of "the position of ontology," particularly the perception of "difference at ontology level," even relates to the transitional development of ontology itself. The "chemical paradigm" regards information as a metaphor; a word that refers to the potential detailed structure of matter/energy. This represents the typical understanding of information as matter/energy. On the other hand, the "information paradigm" regards information as a real fundamental part of the everyday world; yet traces of the old paradigm are obvious as it progresses into the new paradigm, thus the information paradigm has not and even cannot be truly achieved.

Recognizing information as a process is crucial for advancing from understanding it as a relation to understanding it as receptive relation. In this vein, Barbieri gave a great push to the understanding of biological information.

While plants lack independent receptors, the biological information at this level does not directly contribute to understanding information as receptive relation; but there is an important hint from biological information for information understanding: adopting its process nature to grasp the root nature of information. One crucial aspect is the existence of two types of receptivity during its development: the organism receptivity besides the sensory receptivity. It is organism receptivity that can not only be critical to understanding information in plants, but also for understanding the origin of information. Therefore it must involve the problem of chemical or physical reduction in biology, leading to a more proper understanding of information. To this end, Barbieri reached a distinctive conclusion: "Organic information and organic meaning, in conclusion, are not mere names, as the chemical paradigm has claimed: they are fundamental nominable observables" (Barbieri, 2016). Undoubtedly, these efforts have already been made to distinguish information from matter/energy.

However, due to the lack of precise understanding of information, the "information paradigm" cannot effectively win out the "chemical paradigm" in the field of biology. Nevertheless, the "information paradigm" is fundamentally different from the "chemical paradigm," which brings the former totally different purposes and efforts; Baribieri's further exploration is an example of such efforts.

While the indispensability of information is evident, the distinction between this indispensable element and physical variables is not. Merely using the term "sequence" to describe information is obviously not enough, because physical variables have sequence too; only that, in this way, there is no longer any purpose to force the "chemical paradigm" on the "information paradigm." The concept of descriptive entities refers to "abstract entities," i.e., abstract nouns such as "inevitability" or "contingency." This description fully illustrates the non-substantiality and relationality of information.

Although it may be challenging for the "information paradigm" to convince the "chemical paradigm" due to limitations in information understanding, inspiring progress has been made in the understanding of biological information. It is of great importance to realize the difference between information and information encoding, which concerns the critical point for lifting the veil of matter/energy from information. It is important to distinguish between information as receptive relation and information encoding. There are two fundamental forms of information encoding: material encoding of information (e.g., electrical signal and genes) and idealistic encoding of information (e.g., concrete concepts as the reflection of natural kinds and bits as symbols). Digital encoding, as a special form, occupies a unique and significant position.

It is precisely by understanding information as information encoding that the ceiling of information comprehension is established. Regarding material coding of information as information itself is an unavoidable step before achieving a receptive relation understanding of information. It is also the major fog that confuses our information understanding, as in the biology field, as well as in physics.

Barbieri, building on the progress of the "information paradigm", developed his distinctive understanding of biological information and proposed the paradigm for analyzing how information is encoded. Apparently, the "encoding paradigm" captures the essence of biological information on the one hand and holds information in suspension on the other. Since the encoding is information encoding, according to the encoding paradigm, chemical plus information encoding suffices to form this paradigm, only the question of "What is information?" remains unanswered. What can be certain is that the distinction between information and information encoding is a critical step forward for biological information understanding, and this is where humans as receivers can easily get lost, including our understanding of the relation between the material and the spiritual.

The emergence of humans as receivers greatly enriches information, yet simultaneously information gets enmeshed in layers upon layers of complex shrouds. As our ability to understand advances, the improved assigning ability for receptive relation and richer meaning of receptive relation trap us in the complex entangle of information derivatives, akin to silkworms getting enmeshed in a web of their own spinning. Ultimately, we shackle ourselves not only by being unable or unwilling to look at information facts predating the emergence of humanity, but also by being incapable of delving deeply into the relationship between information and matter. As the meanings endowed by receivers during the establishment of receptive relation become more intricate, the level of information complexity increases, and subjective factors appear in the relation and enhance the information subjectivity.

Among the common forms of information that people usually think of, biological information seems to be the most original form of information. However, merely tracing information evolution back to organism is apparently not enough, because information had been earlier generated before organism emerged; otherwise, understanding chemical evolution would be impossible. The matter/energy interaction and information interaction should be seamlessly connected.

2.6 Conclusions

The understanding of information signifies a profound paradigm shift in human cognitive development. In this transformative process, the signal-based information understanding is the first step from matter science especially communication science; the source-based understanding is an important step mainly on the basis of philosophical problems of information science, while receiver-based understanding is a key step mainly for philosophy on the basis of information science and technology development. It is these three approaches and their research outcomes collectively establish the cognitive foundation for transitioning from a matter paradigm to an information paradigm.

In the context of quantum mechanics, coupled with the advancements in big data and AI, there emerges an unprecedented opportunity for the unfolding of information and its receptive relation understanding in this era. Information as receptive relation and its mature state is the dynamic process of receptive relation between source and receiver. There are two kinds of receptivity: sensory receptivity and organismic receptivity. It signifies a genuine shift from a matter paradigm to an information paradigm, which reveals the fundamental characteristics of information distinguishing from matter. This understanding sheds light on the prospects of information in unraveling the mysteries of life, human existence, and the core mechanisms of general AI.

Building upon these characteristics, a series of significant conclusions can be drawn. Research on receptive relation understanding and nature of information holds wide and profound significance for theory and practice. We can see broader prospects for development based on receptive relation understanding of information.

Firstly, the relation paradigm transition of information understanding helps clarify several inevitable misunderstandings about information, and resolves the paradoxical dilemmas encountered in information understanding. The most basic and important paradox is the unbelievable irrational relation between the inevitable information conservation conclusion in physics and the obvious fact that information created ex nihilo.

Secondly, some very important characteristics of information can be drawn from the receptive relation understanding of information such as creating information out of nothing which facilitates our understanding of human creativity and its promising future; and the emergence nature of information as receptive relation which enables understanding holism, reductionism and their relation at information level; the reciprocity of information as receptive relation which provides a new mechanical inspiration for better understanding of anthropology features; and the sharability characteristics of information different from that of matter and energy which offers a new theoretical foundation for a better understanding of human society and its development.

Thirdly, a series of new principles can be introduced by deepening our research based on the receptive relation understanding of information. Among them, the principle of identity for information-matter in operation is the most relevant to the understanding of information.

Finally, by understanding information as receptive relation, we not only deepen our understanding of quantum mechanics and relational ontology but also gain insights into the mysteries of consciousness as well as life. Beyond considering big data merely as a production of digital encoding of information, efforts should be directed toward unveiling the common information basis for AI, its underlying properties, and mechanisms. This approach promises significant advancements in understanding the enigma of the human brain and guides us toward exploring the core mechanisms of AGI in a more appropriate direction.

2.7 Acknowledgment

We are deeply grateful to *Social Sciences in China* for publishing, decisively and quickly, the paper "The Contemporary Interpretation of Information and Its Basic Characteristics" (No.1, 2022).

Note

1 A mountain in China that is always mist-shrouded, so there is a saying: One doesn't know the true Lushan, only because one is in it.

3 The Quantum Manifestation of Information

3.1 Introduction

The receptive relation understanding of information reveals a historic moment in the development of information understanding: upon reflection, one is astonished to witness the quantum phenomenon manifestation of information over a century ago. Quantum phenomena, in fact, are precisely information as receptive relations. However, not only did the quantum manifestation of information go unnoticed for over a century, but "quantum information" was also treated as a mysterious realm, distinct from general information, all because of the material paradigm of information understanding,

Exploring "quantum information" holds significant importance for both quantum physics and information theory, while at the same time, it possesses an air of mystery. The receptive relation understanding of information serves as a clear window, shedding light on the informational nature of quantum phenomena. In the bidirectional cycle mechanism between quantum physics and its manifestation of information, we can also discern the quantum physics presentation of information. In this interplay, "quantum information" holds crucial significance in understanding the concept of information.

The complexity of the information concept becomes even more enigmatic due to "quantum information," and the limitations in understanding the concept of information make understanding quantum mechanics itself even more complex. Many thinkers, including physicists and philosophers, share an important intuition: quantum physics and information are intrinsically connected. In contemporary scientific development, the status of quantum physics is intriguing. On the one hand, quantum mechanics has long been a fundamental pillar of theoretical physics; on the other hand, the theoretical explanation of quantum mechanics has always been foggy. From the mysterious connection between quantum physics and information in quantum information, we can intuitively see both the informational approach to quantum physics and the quantum physics approach to informational understanding, as well as see that quantum mechanics has added several layers of fog. On one hand, "the question 'What is quantum information?' is still far from having an answer on which the whole quantum information community agrees" (Lombardi etc. 2017, p. 1). On the other hand, "Many philosophers and physicists have expressed great hope that quantum information theory will help us understand

DOI: 10.4324/9781003484851-3

the nature of the quantum world. The general problem is that there is no widespread agreement on what quantum information is" (Duwell, 2010, p. 171). As long as the question of "What is information?" remains elusive, delving into "quantum information" undoubtedly adds another layer of complexity, but it also harbors intriguing nuances that demand deeper exploration for both information and quantum physics. In quantum physics where information stands out unprecedentedly, there is still no consensus on "What is quantum information?" which is enough to show the importance and complexity of the concept of information to the understanding of quantum mechanics.

Due to the key position of quantum phenomena in information understanding, understanding quantum information fundamentally involves how well we understand information. As a key link in information comprehension, the understanding of information concepts can only be achieved within a higher-level theoretical framework of science and philosophy. Whether confined to the scientific description of quantum mechanics or its philosophical interpretations, it is difficult to deal with issues related to quantum information understanding which encompasses both scientific and philosophical integrated properties and it is necessary to reach conclusions beyond the boundaries of either science or philosophy.

For humanity, there exists a profoundly significant and mutually enlightening relationship between information concepts and quantum physics. The study of quantum information should have already reflected to varying degrees, conscious or unconscious, this mutual enlightenment, even though it has not been fully demonstrated due to the incomplete understanding of what "quantum information" truly represents. It is precisely within the understanding of quantum information that the mysteries of information concepts and the full understanding of quantum physics intersect.

Understanding the mysterious connection between quantum physics and information is one of the most peculiar situations encountered in the history of human knowledge. On one hand, in the realm of quantum physics, information is prominently displayed because it becomes clear through the receptive relation understanding of information that quantum phenomena themselves are typical examples of information as receptive relations. On the other hand, throughout the course of human cognitive development, opportunities to discover the mysteries of information through their inherent connection with quantum physics have been consistently missed. This leads us to marvel at a historical fact: a thin veil of time can act as an impenetrable curtain under the constraints of history. The reason why this historical fact stands out most prominently in the course of human cognitive development is that it represents a paradigm shift that can be understood based on the receptive relation understanding of information: the span between two paradigms spans vastly different domains, namely, matter and information.

For deepening our understanding of information, research in quantum information provides key insights. The fundamental reason lies in the fact that the informational phenomena in the quantum domain are at least superficially entirely distinct from those in classical physics. Yet it is precisely due to the complexity of quantum phenomena that the concept of "quantum information," with its enigmatic nature, holds the potential for the most promising approach to understanding information.

3.2 Information Stripping in Quantum Observations

In the realm of quantum physics, due to the mismatch between human senses and the scale of objects, the observational context is entirely different from that in classical physics and everyday experience. The starkly contrasting situations between the two in observation become clearer through the receptive relation understanding of information.

Through the receptive relation understanding of information, it becomes apparent that any observation is an informational process of establishing receptive relation. The observed results are all receptive relation effects generated by the interaction between the receiver and the source as well as the experimental arrangement. It's only in macroscopic observations, where human senses match with the objects that observed results as receptive interaction, effects are naturally attributed to the objects, such as the green effect generated by our naked eye observation of leaves, which is considered to be the inherent color of leaves themselves. In observation, the effects generated by receptive interactions can be attributed to the nature of the source of information or the object. In the dichotomy of subject and object in cognition, information as receptive relation will not appear between receiver and source in a way that is neither an objective attribute of the object nor a subjective consciousness of the subject. This is closely related to the characteristics of human senses and the scale of the observed object.

In the macroscopic realm of human observation, whether we use the naked eye or a telescope to observe objects—such as the moon, since the interaction effects of light on the moon can be ignored during the observation process, the effects generated by observation can be regarded as entirely reflecting the nature of the moon. However, if we are dealing with atomic systems, since "If, however, we have to do with atomic systems, whose constitution and reactions to external influence are fundamentally determined by the quantum of action, we are in a quite different position" (Bohr, 1963, p. 11). In quantum theory, matter is considered to be composed of indivisible quantum. In other words, as far as we know, these constructs are the lightest objects. For these objects, any kind of irradiation, or any observation action, will cause an unavoidable disturbance that cannot be ignored; any such disturbance is enough to affect the behavior of the observed object, causing it to undergo decisive changes. Due to this fact, there exists interference in the observation of atomic objects that cannot be treated in the same way as that in the macroscopic domain. This interference cannot be ignored, eliminated, or attributed to the properties of the object.

In observations in the quantum domain, we encounter an important fact: microscopic objects possess quantum properties, and fundamental particles cannot be further divided. Photons, as fundamental particles, also possess quantum characteristics. Like other fundamental particles, a photon is a fundamental unit and does not contain any smaller components. However, observing a microscopic object requires a beam of light or a stream of photons to be directed at it and these photons will impact the observed particle. Since a photon is the most basic unit, this impact force has a certain fundamental quantity and cannot be infinitely reduced. If the observed particles are heavy, like a bacterium or even an atom, the impact of photons on them will not result in observable effects. We can assume that they are not actually affected by the photons, so using a lens to focus the deflected photons can provide an image of the target object, allowing us to observe them. However, if it is a subatomic particle like an electron, the situation of observation is entirely different.

For measurements in the quantum domain, an important characteristic of photons becomes a crucial factor with fundamental impact: its wavelength is inversely proportional to its energy. The longer a photon's wavelength is, the smaller its energy becomes; shorter wavelength means larger energy. Because the photon's "resolution" ability is related to its wavelength, the shorter the wavelength, the stronger the "resolution" ability. When using photons with higher energy for observation, the interference is also greater. Therefore, when we attempt to observe electrons through a microscope, we face an inescapable dilemma.

If we were to observe electrons using a regular microscope, it would be impossible to resolve them under visible light because the wavelength of ordinary light is larger than the diameter of electrons. To observe an object through a microscope, the diameter of the object must be greater than the wavelength of the light used for observation. When the diameter of the observed object is smaller than the wavelength of the observing light, the light will bypass the object without being reflected by it. Therefore, we must use light with a wavelength smaller than the diameter of electrons, such as gamma rays, for observation. However, if we attempt to observe electrons with gamma rays whose wavelengths are smaller than the diameter then another difficulty arises: this is because the wavelength of light is inversely proportional to the momentum of photons. Gamma rays with very short wavelengths have much greater momentum than the original momentum of electrons. When using such light to observe electrons, just one photon can knock the electron out of its orbit, making it disappear without a trace, and thereby destroying the conditions for the existence of the quantum state. As a result, the observer cannot see the electron at all, and in fact, the observation itself becomes impossible.

At this point, the "invisible" barrier between human observers and quantum objects in quantum observations has become faintly discernible: it is a barrier of receptivity lying between human observers and quantum objects. The existence of this receptivity barrier is due to the natural threshold limitations of human senses from the subjective perspective, the quantum properties of microscopic objects from the objective perspective, and the inability to directly establish receptive relation process as information between human observers and quantum objects. In quantum observations, humans can only establish indirect receptive relation through the interaction with quantum objects via experimental arrangements. Therefore, due to different experimental arrangements, different receptive interaction effects – quantum phenomena – can be generated. For instance, in the double-slit experiment, quantum phenomena indicating wave-like behavior – the interference of light – are generated. In the Wilson cloud chamber bubble chamber experiment, quantum phenomena indicating particle-like behavior – the tracks formed by the vaporization of water molecules – are generated. This not only constitutes the paradox of wave-particle duality but also clearly indicates that the quantum phenomena generated by observation cannot be attributed to quantum objects. They cannot be viewed as inherent properties of quantum objects, as is the case in classical physics and everyday observations, nor can observation results be handled in this manner.

Because quantum phenomena as observational effects can no longer be attributed to the properties of objects, there is a separation of information in quantum observations. Information, as receptive relation, can no longer be attributed to the objects or sources of information but exists as receptive interaction effects between the receiver and the source, which is transformed from the inherent properties of the object into information as quantum phenomena. What is highlighted here is the same nature shared by all observations, even when analyzed in the manner of macroscopic observations, which allows us to see this clearly.

In fact, this is the case for any observation, it's just that in classical physics and everyday observations, we can approximate the receptive interaction effects generated during observation as the inherent properties of the object, and this does not affect our understanding and practical needs. Therefore, in the material paradigm, the understanding of quantum information, quantum physics becomes the most fundamental field of the source-preexisting understanding of information. However, if we consider quantum observations on the basis of understanding information as receptive relation, we can see the magical connection between quantum physics and information.

The magic of quantum mechanics in understanding information lies in the direct manifestation of information through quantum phenomena.

3.3 Quantum Phenomenon Itself Is Information as Receptive Relation

In quantum observations, we not only directly engage with information as receptive relation but also encounter an intriguing phenomenon: as a more complex receptive relation, the quantum phenomena generated by quantum observations not only do not make the understanding of information more difficult but actually highlight information as receptive relation. Therefore, quantum information not only doesn't complicate the understanding of information, but quite the opposite; the receptive interaction effects generated by quantum observations provide an unprecedented and clear manifestation of information.

It is precisely the indirect nature of observations in the quantum domain that emphasizes that what we observe is not the object itself but the effects of the interaction between experimental arrangements, including the observer and the object – these are quantum phenomena. Therefore, Niels Bohr believed that the concept of "phenomenon" in quantum physics needed to be redefined.

Bohr discovered that even the term "phenomenon" could not be used without introducing new meanings in quantum theory. In classical physics, it seems relatively easy to distinguish between subjective factors and objective phenomena.

Objective reality appears to be imposed in the same way by the external world on all people; any observation of it can disregard the thoughts and actions of the observer. In other words, observers can naturally talk about objective phenomena at any time, and human senses serve as somewhat imperfect tools that help us gain knowledge about the objective world. Therefore, improving our senses through artificial observation methods, aiming to delve into the farthest realms of objective reality beyond direct receptive range, seemed natural and logical for physicists. However, it is precisely here that what Heisenberg called the "hope of deception" arises, which suggests that through further improvements in observational methods, we might eventually be able to understand the entire world.

The idea that one can study an object without disturbing it is evidently an illusion of classical physics. In fact,

> our experiments are not nature itself, but a nature changed and transformed by our activity in the course of research. To effect a real change would undoubtedly entail a complete abandonment of the whole of modern technology and science, which is linked with it.
>
> (Heisenberg, 1979, p. 71)

There is no doubt that an observable quantity is not something in itself but a property of a relation of it with other things. This is true even in classical physics, but people can approximate the nature of the relationships between things as an inherent property of one thing. In classical physics, the distinction between the independent action of the object and the means of observation is assumed to be possible with respect to the goal of human understanding. All empirical descriptions are built on this inherent assumption of common linguistic convention. This convention is not only fully confirmed by all everyday experiences but also forms the entire foundation of classical physics. However, in quantum physics, because all clear knowledge about quantum objects is derived from permanent marks left on macroscopic instruments during observation we face an entirely new situation.

As Bohr pointed out,

> As soon as we are dealing, however, with phenomena like individual atomic process which, due to their very nature, are essentially determined by the interaction between the objects in question and the measuring instrument necessary for the definition of the experimental arrangement, we are, therefore, forced to examine more closely the question of what kind of knowledge can be obtained concerning the objects.
>
> (Bohr, 1958, p. 25)

In the physics of macroscopic systems, events are represented as sequences of states described by measurable quantities. This approach is based on the fact that all quantities involved have such large action that we can neglect the interaction between the object and the means used as measuring tools. When action quantum plays a decisive role, making this interaction an indivisible part of the phenomenon, it is impossible to attribute events to a mechanically well-defined procedure to the same extent. Precisely because of this, Bohr pointed out that the interpretation of the paradoxes of atomic physics reveals the fact that the inevitable interaction between the object and the measuring instruments places an absolute limit on the possibility of speaking about some behavior of the atomic object that is independent of the means of observation. This limitation is what restricts our discourse on objective objects.

In fact, the discovery of the role of quantum mechanics not only revealed the natural limitations of classical physics but also puts us in an unprecedented situation in the field of natural science. This situation is revealed through a reexamination of an age-old philosophical question: the objective existence of phenomena is not dependent on our observations. Since any observational result concerning quantum objects ultimately arises from their interaction with the entire experimental arrangement "The limit, which nature herself has thus imposed upon us, of the possibility of speaking about phenomena as existing objectively finds its expression, as far as we can judge, just in the formulation of quantum mechanics" (Bohr, 1932, p. 115). The facts revealed by quantum theory immediately turned this "hope of deception" into bubbles. Any doubt about the existence of objects independent of any observer is contrary to human reason.

This leads to a profound philosophical question: if the act of observation constitutes an uncontrollable interference with the object, if the presentation of the objective world depends on the behavior of the observer, then it's very difficult for us to talk about a purely objective state of the world, one that is unrelated to observation. Thus, the concept of "phenomena" in quantum physics must be redefined.

In this sense, Bohr believed that in the quantum domain, the term "phenomena" refers specifically to the results of observations within the observer's domain and can no longer refer to the state of things that exist independently of the observer, as is the case in the macroscopic realm. In other words, "quantum phenomena" do not refer to properties such as the external connections and surface features of electrons themselves, but rather to their tracks in a Wilson cloud chamber or the permanent marks they leave when passing through photographic emulsion. Bohr wrote:

As a more appropriate way of expression I advocated the application of the word phenomenon exclusively to refer to the observations obtained under specified circumstances, including an account of the whole experimental arrangement. In such terminology, the observational problem is free of any special intricacy since, in actual experiments, all observations are expressed by unambiguous statements referring, for instance, to the registration of the point at which an electron arrives at a photographic plate. Moreover, speaking in such a way is just suited to emphasize that the appropriate physical interpretation of the symbolic quantum-mechanical formalism amounts only to predictions, of determinate or statistical character, pertaining to individual phenomena appearing under conditions defined by classical physical concepts.

(Bohr, 1958, p. 64)

From this, Bohr drew the significant conclusion: "On the line of objective description, it is indeed more appropriate to use the word phenomenon to refer only to observation obtained under circumstance whose description includes an account of the whole experimental arrangement" (Bohr, 1958, p. 73). In essence, the quantum "phenomena" that Bohr sought to redefine is information as receptive relation, it's just that here information is highlighted in the quantum domain. Moreover, quantum phenomena that appear to be incompatible with human sensory characteristics are essentially no different from phenomena that are compatible with human sensory characteristics. The difference lies in the fact that they are not external manifestations of thing-in-itself as they appear in the Kantian sense but direct manifestations of information as receptive relation. The definition of quantum phenomena as receptive relation makes sense only at the level of information.

The receptive relation understanding of information makes it clear that it is through instruments and experimental arrangements that human observers are able to perceive microscopic objects, and presents further that what we get is merely the effects – receptive relation – of the interaction between receiver (observer) and experimental arrangement with source (the external objects); because of this precisely leads to an important discovery: quantum phenomena are information itself as receptive relation!

The receptive relation understanding of information unveils the mystery of quantum mechanics at the quantum level: quantum phenomena are the marvelous manifestation of information itself. The mechanism involves the special relationship between receiver as observer and the source as quantum objects in quantum observations, stripping receptive interaction effects attributed to the inherent properties of objects in classical physics and everyday empirical observations, and directly presenting information as receptive relation. From presentation of information in quantum phenomenon looking back at phenomena in classical physics observation and daily experience, their nature is exactly the same it's just that due to observation effect being attributed to the object's inherent properties we have always missed understanding of information as receptive relation.

In the daily life domain, considering the "green" effect generated by our visual system with light reflected by leaves as inherent property or even pure objective existence of the leaves can be described as "one leaf blinds the eye" or a "blind spot." In contemporary science, the relational nature of color effects has become clear. Therefore, what we see with our eyes, whether "green" presented by leaves in our eyes or quantum phenomena; they are all information itself as receptive relation. From this perspective, not only is there a typical manifestation of information, but there is also a promising future for understanding information in quantum physics. Clearly stating that quantum phenomenon is information itself as receptive relation not only for understanding information but also for deepening understanding of quantum physics has great significance.

3.4 Information Understanding of Quantum Physics

The informational nature of quantum phenomena indicates that quantum physics deals with facts that are distinct from those in classical physics. To put it more precisely, quantum facts more accurately represent the nature of observed "facts." Therefore, understanding quantum phenomena at the level of information as receptive relation is of significant importance, both for the understanding of information and the deepening comprehension of quantum mechanics.

3.4.1 Information Seeking for the Physical Meaning of Quantum Mechanics

Due to the special situation of human senses in quantum measurements, quantum physics reveals that the nature of informational facts is different from that of material facts. One of the most fundamental distinctions involves the objectivity of informational facts. Information facts do have properties different from material facts, with one of the most significant differences being the absence of the strong objectivity found in material facts. It is precisely because of this that informational facts aid in understanding and explaining the epistemological challenges brought about by pure objectivity, such as the issues of locality and non-locality. Here, the understanding of informational facts is crucial, as it relates to the mutual insights of quantum physics and information theory, involving scientific and philosophical implications of the concept of information.

Without a comprehensive understanding of information, it is impossible to have a fundamental grasp of the essence of information, let alone truly understand quantum physics. The longstanding divergence of quantum mechanics and its theoretical interpretations has told us that seeking the physical meaning of quantum mechanics in the classical physics sense, or more precisely, limited to the paradigm of materiality, is no longer feasible. It has even lost its significance, as the physical meaning itself is a product of the material paradigm. Seeking the physical meaning of quantum mechanics within the material paradigm is destined to be constrained by that paradigm. What we can and should seek regarding quantum physics is to understand its physical meaning through the lens of the information it conveys, essentially seeking the information-physical meaning of quantum mechanics. This, in itself, not only holds rich scientific and philosophical implications but also embodies significant integration of science and philosophy. This is precisely the characteristic of quantum information research itself and its research content. Indeed, different information concepts can be used for different purposes, but this is only said in terms of use. From this, it cannot be concluded that there is no unified answer to the question of what information is. Because this means that we cannot elevate our understanding of information to a level higher than the use context, which is not only based on traditional metaphysical criticism, but also obviously not true in terms of the holistic guan zhao of theory.

Just as with the use of tools, although there is much flexibility in the use of concepts, there is still a question of reasonableness. From a formal perspective, different understandings of the concept of information have their specific meanings, and the key here is to place the understanding of the concept of information within both empirical and formal relational patterns.

Returning to Timpson, while explicitly treating information as an abstract noun, one can see the significant work that analytical tradition can do for this. However, if it stops there, the understanding of information will not be commensurate with its status alongside matter and energy. In fact, the concept of information is not like any abstract noun in mathematics; it implies an important empirical theory. Furthermore, information theory is not just a discipline parallel to communication science and statistics; it is a deeper research field. Due to the particularity of quantum mechanics, quantum information research involves the clarification of the concept of information at a deeper level. In a recent overview of quantum information research, Olimpia Lombardi et al. highlight this point:

> "the question about what quantum information is can be addressed from many different perspectives, some of them complementary, others conflicting with each other. This plurality precisely reveals to what extent the community of quantum physicists and philosophers of physics is far from a consensus about the answer of that question." For this reason, they were dedicated to discuss "about the conceptual foundations of an exciting field like quantum information theory."
>
> (Lombardi et al. 2017, p. 5)

To avoid confusion about the concept of information, in addition to artificially limiting it to a certain meaning, a higher level of theoretical holistic guan zhao can also be raised. Even though this may not necessarily lead to a more accurate description, it provides an indispensable approach to approximating this goal.

Indeed, "information" is an abstract concept; it is an abstract generalization that provides holistic guan zhao for our concrete understanding of receptive relation. However, the connotation it reflects is concrete receptive relation. Therefore, what "information" reflects as an abstract concept is actually concrete receptive relation that serves as its connotations, in other words, information as concrete receptive relation. Consequently, at a deeper level of information, there appears to be a shift in fundamental philosophical perspective: from the ultimate pursuit of abstract universality to the abstract universality that only provides holistic guan zhao for the understanding of concrete individual things. In the understanding between quantum physics and information theory, one can observe this reciprocity of holistic guan zhao.

3.4.2 The Informational Root of Quantum Paradoxes

From the receptive relation nature of quantum phenomena, one can derive an informational interpretation of quantum physics. Furthermore, based on the receptive relation understanding of information, quantum paradoxes can be more reasonably comprehended.

Since quantum phenomena are information itself as receptive relation, quantum physics should be a physics that further considers information on the basis of quantum mechanics as a mathematical formal system. Observations in the quantum field show that the construction of quantum theory must incorporate receptive relation in that the observer is receiver. This is an inherent characteristic of human observation, but the nature of receptive relation is typically presented in quantum phenomena. In the understanding of receptive relation of information, whether it is waves or particles, quantum phenomena generated by quantum observations are not inherent properties of quantum objects themselves but are instead properties of quantum phenomena as receptive relation.

Therefore, about "quantum paradoxes" derived from this perspective, we can get a more reasonable information understanding: as a property of the object itself, the quantum object cannot simultaneously be both waves and particles as inherent properties, whereas as information as receptive relation, there is no conflict between the two presentations of quantum phenomena at all. This means the wave-particle paradox in quantum theory only exists in macroscopic experiences. With the information technology revelation of information, the deepening understanding of information not only reveals quantum phenomena but also reveals the informational nature of all observations, including our daily observations.

The receptive relation understanding of information shows that what is obtained from observation is only receptive relation generated by interaction of the receiver as observer with the source as quantum object which is neither an objective inherent property of the source nor purely the source itself. Just as seeing a leaf as green is, in reality, a receptive effect resulting from the interaction of the visual system with the light of a specific wavelength, approximately 600 nanometers, reflected by the leaf, the green effect is not a property of the leaf itself but merely a receptive effect process resulting from the interaction of the human visual system with the specific wavelength of light reflected by the leaf. This specific "green" effect, as a receptive effect, is neither the objective property of the leaf nor a projection purely originating from the receiver but is information itself as receptive relation.

Through the analysis of these everyday phenomena, based on the receptive relation understanding of information, a more reasonable interpretation of seemingly inexplicable and mysterious quantum phenomena, such as the "Schrödinger's cat" and even the "quantum Cheshire cat," which are typical examples, can be provided.

As a thought experiment used to challenge the concept of quantum superposition, "Schrödinger's Cat" is pivotal, as it involves receptive relation as information. In the thought experiment, a cat is placed inside a box, and there is a 50% chance that a device inside the box will release poison gas. According to the interpretation of the principles of quantum mechanics, before the box is opened and the cat is observed, the cat exists in a superposition of being both alive and dead simultaneously. This is what's known as a quantum superposition state, meaning the cat is both alive and dead at the same time. Only when the box is opened and the cat is observed does the wave function collapse, and the cat's state returns to either being alive or dead. In the framework of material paradigms, "Schrödinger's Cat" is incredible, but when viewed from the perspective of information as receptive relation, it's about the probability description of the actual state of the cat or a statistical or mathematical description of the real world it entails. In a real-world scenario, if a transparent box were used in the experiment, allowing us to continuously watch the cat, there would be no real wave function or its collapse. Here, the crucial point lies in the fact that the barrier of sensitivity to quantum observations becomes the basis for the impossibility of the box being transparent. However, as a thought experiment, it can be understood that "watching" means establishing receptive relation process, which in turn implies information. In other words, the reason "Schrödinger's Cat" is in a superposition of neither dead nor alive is not because the cat genuinely exists in such an incredible actual existence state but because when receptive relation as information was not established or could not be established probability description for actual existence state of the cat. From this, one can grasp the descriptive nature of quantum superposition through the macroscopic state of "Schrödinger's Cat." At the level of information as receptive relation, "Schrödinger's Cat" not only becomes understandable but also leads to some important conclusions. One of the most obvious is that the receiver understands the source must be based on receptive relation as information, while under conditions where corresponding receptive relation has not been or cannot be established, our understanding of the source can only be achieved through mathematical descriptions or conceptual speculation. The difference between "Schrödinger's cat" thought experiment and quantum phenomena is primarily that quantum transitions imply an inability to establish receptive relation with the process, which is not a problem in the macroscopic domain. According to the receptive relation understanding of information, a similar interpretation can be made for the "quantum Cheshire Cat."

Another thought experiment, the "quantum Cheshire Cat," is used to illustrate the paradoxical nature of quantum mechanics. In quantum physics, a photon's circular polarization can be separated, much like the legendary "Cheshire Cat" whose smile can be separated from its body. In this thought experiment, a photon is analogous to the Cheshire cat, and its circular polarization is analogous to the cat's smile. In the framework of the material paradigm the Cheshire Cat whose smile can separate from its body is also incredible and it is incomprehensible that the performance of a cat can exist separately from its physical entity. The physical properties of a photon can be separated from its physical entity, meaning the properties of a thing can be separated from the thing itself, which cannot be explained or understood within a material paradigm because, as physical explanation, it implies that any quantum physical property can be separated from its physical entity, or any quantum entity can be separated from its physical properties. However, in the information paradigm, this phenomenon can not only be understood but also clearly indicates a deeper understanding of the relationship between things and their properties. In the receptive relation understanding of information, the so-called properties of the source, as the object of receiver perceives, are all manifestations of information as receptive interaction effects. In this sense, the information understanding of the "quantum Cheshire Cat" suggests that it's not that the properties of source can be separated from its physical entity; instead, the so-called properties of the source are not inherent qualities but rather receptive interaction effects between the receiver and the source.

In quantum physics and classical physics as well as everyday experience, the so-called properties of things, are, in reality, information as receptive interaction effects. In the thought experiment of the "quantum Cheshire Cat," the smile or circular polarization is all quantum phenomena as information as receptive relation. Through some of the stars we can see in the night sky, this understanding can be further elucidated.

Some of the stars we see in the night sky have actually been annihilated many light years ago, yet we can still see the light they emitted and form the receptive interaction effect of seeing them today because, as receivers, we have established receptive relation with the light they emitted before they were annihilated, which allows us to infer that they still exist now. In this scenario, observation effect as receptive relation effect, what we see (stars) are the smiles of Cheshire cats separated from Cheshire cats themselves – stars, like the Cheshire cat's smile, exist when the physical stars, like the Cheshire cats, themselves no longer do! Clearly, this not only provides a reasonable explanation of the "Quantum Cheshire Cat" but also further highlights the importance of information in interpretation of quantum mechanics.

3.4.3 New Approach to Information Quantum Theory

The fact that quantum phenomenon itself is information as receptive relation allows us to see the important connection between philosophical and scientific integrated information research and quantum physics, and to see the informational nature of quantum physics. The philosophical research on quantum physics is intrinsically linked to information theory, representing a significant connection between these two research domains and offering a breakthrough in both philosophical inquiry into quantum physics and the further deepening of information theory research.

In the quantum realm, observations extend beyond human sensory thresholds and manifest indirectly through a mode of indirect recept, highlighting receptive relations and consequently, information. In observations that align with the scales matching human sensory characteristics, we tend to approximate observation as objectively distinguishable, leading to a clear distinction between the observer and the observed object. This illusion makes us believe that what we observe is the object itself, so it feels absurd not to think so. For instance, it seems absurd to assert that what you are reading now is not this actual book being read, and even more so to say: "The book only exists when I look at it." However, at a higher level of information, this understanding is not difficult to grasp; the receptive relation we establish while reading a book is obviously the effect of "seeing" by the reader himself or herself. It's just that in the quantum realm, this understanding can delve deeper into more fundamental layers.

In quantum observations, because we can no longer assume that we can directly observe the object itself, the situation has fundamentally reversed. It is precisely because the observation in the quantum field shows that what is observed is the receptive interaction effect established between the observer as a receiver and the quantum source, so from this receptive relation or information basis, the mathematical formalism of quantum mechanics doesn't manifest physical meaning on its own and necessitates additional theoretical explanations. In classical physics and everyday life, because we believe that what we see is the object itself, we feel that the object is clear at a glance and does not need additional theoretical interpretations.

The clear presentation of information in the field of quantum physics is not easily discovered due to its specificity, and this further highlights the historical limitations of human knowledge development. On one hand, due to its highly specialized nature and being within the understanding in the material paradigm, few individuals can comprehensively grasp the fundamental concepts of quantum mechanics. On the other hand, owing to the multitude of mysteries and challenges existing in the quantum domain, experts are mostly engrossed in exploring quantum mechanics and the profound scientific and philosophical questions it encompasses. Moreover, the mechanism for clear presentation of information in the quantum field is precise indirectness of the observer's recept in the quantum field which makes quantum mechanics only able to establish mathematical formalism based on probability quantification description of quantum phenomena. Because quantum mechanics itself presents as mathematical formal system it masks channel leading to understanding of information as receptive relation; because of masking by physical material paradigm theoretical interpretation of quantum mechanics yield some "bizarre" phenomena. The information understanding of quantum physics not only aids in comprehending these "bizarre" phenomena in depth but also provides new approach for a deeper understanding of the relationship between classical physics and quantum physics.

Regarding the information understanding of quantum mechanics, it is closely linked to the unique association with information of quantum physics. "Quantum theory is about quantum information" (Laura, 2018). If we thoroughly integrate the understanding of quantum mechanics with information, we can even arrive at a genuine theory of informational quantum physics.

3.5 The Information Boundary between Classical Physics and Quantum Physics

From the special scenarios of quantum observations, it is evident that the distinction between classical physics and quantum physics on the macro-micro scale not only is it not fundamental but sometimes may be inappropriate or even misleading. It is this distinction that has kept our understanding confined to the material paradigm. Therefore, it makes sense to say that quantum entanglement and even non-locality is not reasonable based on the boundary of the macro-micro scale.

In fact, in relation to information, quantum mechanics and classical physics do not have a fundamental difference. The differences between them are actually closely related to the attributes of humans as observers. The physical phenomena in classical physics are also informational in nature, but due to the effects of physical interactions, the receptivity of humans as receivers can be completely ignored. The informational nature of the interaction can be seen as purely physical interactions, which leads to the neglect of information because it is attributed to the inherent properties of objects. In quantum physics, however, the effect of the receptivity of observer as receivers, just as the observer's foothold in the theory of relativity, cannot be ignored, and the informational nature becomes prominent as a result. What is highlighted is not only that quantum phenomena are information itself as receptive relation, but also includes all phenomena as observational effects. Due to the fact that quantum phenomena are information itself, the systematic research on quantum information presents a peculiar cognitive landscape where quantum physics and information are keys to each other; and extending the relationship between quantum phenomena and information to all observational effects unveils a deeper common background between quantum physics and classical physics.

In the receptive relation understanding of information, the relationship between quantum physics and classical physics takes on a deeper informational background. In the background of information as receptive relation, whether it is generated by human observers observing objects of different scales or not can be ignored. In the field of classical physics, just as in daily life, human senses match with observed objects, so the phenomena generated by observation can naturally be attributed to the inherent properties of the objects. In the material paradigm, doing so not only poses no problem but is also natural for human being. In the field of quantum physics, however, human senses cannot directly establish receptive relation with quantum objects but must generate quantum phenomena as receptive relation through indirect interactions by experimental arrangements. Unlike phenomena in classical physics and everyday life, quantum phenomena cannot be simply attributed to the inherent properties of objects in a subject-object dichotomy cognition framework, so they cannot be ignored in the establishment of the corresponding knowledge system. Instead, they are the only direct objects for understanding quantum objects. It is in the non-classical field of quantum physics that the key role of information in the theory of knowledge is typified.

Human understanding of quantum objects can only be achieved through information as quantum phenomena. Accordingly, various interpretations of quantum theory can arise based on logical relation, as the logical space of possibilities is greater than the empirical space of possibilities.

From this perspective, the boundary between classical physics and quantum physics essentially lies in whether information, as receptive relation, can be ignored. Since object scales match human senses in classical physics, receptive relation as information can be attributed to objects. This naturally allows the construction of classical physics where receptive relations as information can be ignored, leading to a simple and convenient subject-object dichotomous framework. In the field of quantum physics, because quantum phenomena generated by observation cannot simply be attributed to the properties of the object itself, they can only be interpreted through conceptual systems to achieve a physical understanding of the mathematical formalism of quantum mechanics. Here, an important transformation occurs: the process of establishing receptive relations as information differentiates from the understanding of the receiver based on idea-tional systems. This results in a differentiation in the description and interpreta-tion of quantum theory. Therefore, the establishment of quantum theory clearly consists of two stages: first, the quantitative grasp of quantum mechanics as a mathematical formalism can only be achieved through receptive relations, and then theoretical interpretation is carried out on this basis. That is to say, an indi-rect interpretation can be developed based on this foundation. In other words, the understanding of the quantum world must be carried out indirectly according to quantum phenomena, so not only can we not ignore the quantum phenom-ena as information, but also theoretical interpretation of quantum mechanics as a mathematical formalism can only be carried out based on information as quantum phenomena.

This not only provides an approach to understanding the "explanatory gap" problem but also explains the differentiation in the interpretation and description of contemporary science at the level of information. In classical physics theory, where objects match human senses, the description and explanation have not yet differentiated. Even abstract mathematical descriptions like "F = ma" have direct physical meaning, implying a corresponding physical picture. The important dis-tinction between classical physics and quantum physics is whether a corresponding physical picture can be directly established. In classical physics, the description of objects and their explanation are in a primitive undifferentiated state; they are intertwined. In quantum physics, due to the necessity to grasp the mathematical formalism through receptive relations and then, based on quantum phenomena as receptive relations (establishing ideational systems based on information encod-ing), to interpret quantum objects and build a physical picture of the quantum world, the description and explanation of quantum objects have already differenti-ated. This is the reason for the appearance of phenomena such as the missing of explanation and the loss of understanding in quantum theory. From this, it can be seen that when objects match human senses, allowing the ignoring of infor-mation as receptive relation, there will be no differentiation between description and explanation in scientific theory. However, when information as receptive rela-tion cannot be ignored, and a corresponding ideational system must be established to interpret the object world based on information encoding, the differentiation between description and explanation in scientific theory becomes inevitable. This is a significant difference between classical physics and quantum physics, which further deepens the understanding of their boundaries.

Compared to the physical boundary between classical physics and quantum physics, a deeper-level difference is the information boundary between them: clas-sical physics is a knowledge system that can be established by attributing receptive relation to objects and ignoring information, while quantum physics is a knowl-edge system that cannot be established by attributing receptive relation to objects and ignoring information anymore. It's just that this difference is not consciously made but a natural result closely related to human sensory characteristics. From here we can see epistemological implications of this differentiation: the current "knowledge" is an expression with a "human" default, and the emergence of "machine knowledge" in the development of artificial intelligence demonstrates that the theory of knowledge research into knowledge at the information level increasingly involves the anthropological characteristics of knowledge. The information boundary between classical physics and quantum physics is closely related to this.

From the level of receptive relations, it can be seen that classical physics and quantum physics have an important difference in paradigm levels. Regarding the nature of receptive relation in quantum phenomena, quantum physics can be considered information physics, but strictly speaking, quantum mechanics is not yet exactly "quantum physics," it is more appropriately called mechanics because its nature is entirely different from classical physics, with similarities only in the formulaic grasp of interaction processes. To truly be called quantum physics, quantum mechanics must have an understanding of its physical meaning. However, in this regard, the current interpretations of quantum theory are still not entirely satisfactory. On the basis of understanding quantum phenomena as information, it is possible to further elevate quantum mechanics as a genuine quantum physics based on its mathematical formalism.

From the information level, classical physics can be seen as physics established in the material paradigm, while quantum physics can be seen as physics established in a higher-level information paradigm. In this sense, the former can be understood as "matter physics" (although a bit weird, that's the product of the intersection of two fundamentally different paradigms), and the latter is true "information physics." Based on information as receptive relation, the understanding of "information physics" provides a deeper foundation for the clarification of the "quantum information" haze.

3.6 Clarification of the Fog around "Quantum Information"

In current understanding, "quantum information" signifies a more fundamental information theory.

> Quantum information is a rich theory that seeks to describe and make use of the distinctive possibilities for information processing and communication that quantum systems provide. What draws the discipline together is the recognition that far from quantum behavior presenting a potential nuisance for computation and information transmission (in light of the trend toward increasing miniaturization) the fact that the properties of quantum systems differ so markedly from those of classical objects actually provides opportunities for interesting new communication protocols and forms of information processing. Entanglement and non-commutativity, two essentially quantum features, can be used.
>
> (Timpson, 2013, p. 45)

Combining quantum and information into a field of quantum information is indeed related to the distinct characteristics of quantum systems compared to classical physical systems, but fundamentally stems from the receptive relation of quantum phenomena itself.

As a reflection of a new research area, the concept of "quantum information" holds significant importance in and of itself: an understanding and study of information within quantum physics. Quantum information is a burgeoning field of research, not solely due to the importance of quantum communication and quantum computing, but also because it is deeply intertwined with both the interpretation of quantum mechanics and the understanding of information. Timpson has conducted systematic research on "quantum information," providing a clear elucidation of the subject at the disciplinary level. He states,

> Quantum information is a field at the intersection of quantum physics, communication theory and computer science. It has considerably increased our understanding of quantum mechanics, developed our conception of the nature of computation, and spurred impressive increases in our ability to manipulate and control individual quantum systems. Not only that, but the theory hints enticingly at ideas of rich philosophical promise.
>
> (Timpson, 2013, p. vii)

Timpson's noteworthy feature in quantum information research is that the entire research is based on an analytical tradition. He argues that the everyday concept of information should be strictly distinguished from the technical concept of information theory, but in both cases, "'information' functions as an abstract noun, hence does not refer to a particular or substance" (Timpson, 2004, Abstract). Therefore, by distinguishing the everyday concept of information from the technical concept of information theory, Timpson arrives at an important point in the understanding of information: "'information' is an abstract noun" (Timpson, 2013, p. 3). Just as "matter" is an abstract noun, "information" is indeed an abstract noun; as an abstract noun, "information" is an abstract generalization of specific individual things, just as "matter" generalizes another class of specific individual things. However, the abstract nature of "information" as an abstract noun has important peculiarities, at least logically. It naturally leads to the conclusion that information does not refer to something specific or concrete. This is not only some kind of negation about the signal-preexisting understanding, source-carrying understanding, and receiver-assigning understanding of information, but logically implies a more reasonable direction for understanding information. However, it's not easy to conclude that "information" is an abstract noun on one hand and obviously not enough on the other hand. It still needs to go deeper into understanding about information. Information does not refer to "a specific" thing or "entity" leading to the conclusion that "information" is an abstract noun and thus it could be that "information" belongs to non-entity existence. The receptive relation understanding of information suggests that what the concept of information refers to may not be the relation between specific things or entities. Therefore, from Timpson's systematic research on quantum information, one can see rich and important intellectual resources in this regard.

Regarding the understanding of information, the concept of "quantum information" is indeed insightful, but the specificity of "quantum information" actually reflects the uniqueness of the relationship between receiver and source in quantum observations. This involves the complex relationship between the signal and the physical carrier (medium) of the signal and its characteristics during the observation process. The uniqueness of the signal carrier certainly falls within the scope of information research, but it is not the direct task of understanding information itself.

Observation involves the physical carriers (mediums) of signals, with light being the most typical; information processing involves signals, most typically electric pulses. Both have the same function but completely different natures, and they differ in whether they involve receptive relation. In communication science, electrical signals are typical of material information encoding. Research and processing of electrical signals do not directly involve information as receptive relation in communication science. However, in quantum physics, because quantum observation is about establishing receptive relation, it directly involves information. This is precisely why, in contrast to communication science, people think that quantum mechanics involves a special form of "quantum information."

Due to the complex relationship among objects, signal carriers (mediums), and human visual receptivity, receptive relations generated by quantum observations are far more complex than those generated by classical physics and everyday life observations. However, this does not imply the existence of a completely different "quantum information" separate from general information. The complexity of information in quantum observations lies not in the difference in information itself, but in the distinction arising from receptive relation originating from different signal carriers.

In this sense, "quantum information" is not a different type of information, nor is it more mysterious than information in the general sense. Even the so-called "intrinsic randomness" is not unique to quantum information. Olimpia Lombardi et al. argue that " intrinsic randomness marks the difference between quantum and classical information: quantum information is a type of information that is only possible in a world in which there are intrinsically random events" (Lombardi etc., 2017, p. 3). In fact, intrinsic randomness is not limited to the quantum level but is a property of all process relationships. This means that as receptive relation, information invariably possesses intrinsic randomness, but it becomes more negligible in interactions primarily governed by the physical properties of matter, such as in classical physics; while in interactions that are mainly informational in nature – typically like quantum physics – inherent randomness cannot be ignored. Because fundamentally speaking, inherent randomness stems from the uniqueness of specific things. Even if particles can be considered "identical" spatially during interaction processes, they can be entirely distinct in the dimension of time. Since information has a nature that cannot be divorced from specific contexts and the flow of time, it cannot be treated as spatially and temporally independent (to be precise it can be abstracted using space-time stipulation), and thus, it cannot be easily generalized. For instance, attempting a dichotomy abstraction for multi-information agent factor informational interactions is almost impossible. In essence, the so-called "intrinsic randomness" is not the nature of things but characteristic of the agent's description of things. Clarifying this point is fundamentally significant for the understanding of "quantum information."

Current research in "information quantum mechanics" primarily pertains to tasks within quantum physics, rather than focusing on information itself, as a consensus on quantum information is still in progress. However, this research is inherently connected to the understanding of information, and starting from information, quantum information indeed serves as a key to deepening our understanding of quantum physics.

To many, quantum information may seem more deeply hidden than classical information. In reality, it is quite the opposite. It is the indirect nature of observations in the quantum realm that actually allows information to be presented more clearly. Therefore, rather than saying that quantum information veils deeper mysteries of information, it implies a deeper understanding of the concept of information. One could even say that quantum information represents a more direct presentation of the fundamental concepts of information than in other contexts. In this sense, considering "quantum information" as a separate field of study or even a discipline is, at least in terms of understanding, a misconception. Its roots lie in the physical understanding of information, and its essence lies in the study of quantum-level signal carriers.

3.7 Conclusions

In quantum observations, since human senses cannot directly perceive quantum objects, observation must generate quantum phenomena as observational effects through experimental arrangements, providing an indirect understanding of quantum objects. Due to the unique relationship between light as the medium of observation and quantum objects in the quantum domain, the receptive interactions in observation can no longer be attributed to the properties of the objects themselves, resulting in a detachment from the objects. It is this detachment that allows information, as receptive relation, to be directly presented. Over a century ago, quantum physics vividly presented information as receptive relation; quantum phenomena themselves are the information manifested as receptive relation. The direct cognition object is the information as receptive relation between the receiver as observer and the quantum object. As a quantum observational effect, the quantum phenomena generated by observation are not inherent properties of the observed objects but rather receptive relation generated by the observation. Quantum phenomena, as a quintessential form of information, are, in fact, marvelous presentations of information. In reality, not only in the quantum field, all observational effects are information as receptive relation.

This can lead to a mutual interpretation of information and quantum physics. On one hand, quantum physics directly presents information as receptive relation through quantum phenomena; on the other hand, the receptive relation understanding of information deepens our understanding of quantum physics to an informational level and provides a new foundation for seeking the physical meaning of quantum mechanics at an informational level. Whether "Schrödinger's Cat" or "Quantum Cheshire Cat," as quantum paradoxes, both is closely related to the receptive relation nature of quantum phenomena. Without establishing receptive relation, "Schrödinger's cat" can only be described as miraculously both dead and alive; because quantum phenomena, as receptive interaction effects in quantum observation, are not inherent properties of quantum objects, the separation of the smile and the cat's body in the "quantum Cheshire Cat" scenario is not only conceivable but is its true nature. Through the receptive relation understanding of information, it is even possible to see the path toward the development of a truly meaningful theory of quantum information.

Furthermore, this perspective allows us to clarify the deeper distinctions between quantum physics and classical physics at the level of information and dispel the mystique surrounding "quantum information." Classical physics is a kind of theoretical system established in a situation that can ignore information by attributing receptive relation to object properties; whereas quantum physics is established in the situation that because it can no longer ignore information by attributing receptive relation to object properties. Similarly, as receptive relation, "quantum information" is not a deeper level of information different from general information.

The complexity and apparent mystery of so-called "quantum information" do not stem from the information itself within the field of quantum physics but from the distinctive relationship between observers and information in quantum observations, which differs from the macroscopic domain. To some extent, the nature of information is the same in all domains, and the complexity and mystique around information inherent in natural sciences arise from the unique properties of the relationships between observers and information in these domains, with the most significant factor being the signal carriers that serve as the medium of observation.

Based on the receptive relation understanding of information, the clear quantum phenomena presentation of information is so striking and typical, but for over a century, we have overlooked it! This has nothing to do with human ability; the long neglect of the quantum phenomena presentation of information is closely related to the inevitable historical limitations of human cognition development. In the research on information within the material paradigm, the masking of information as receptive relation is a significant factor.

Due to its unique nature and high specialization, the clear presentation of information in the field of quantum mechanics is easily overlooked. Due to indirectness in observer recept in quantum observation and detachment in description and explanation that makes quantum mechanics only presentable in mathematical formal formalism, it still needs to explain its physical meaning. Even though a historically significant major turning point in physics is implied here because thanks to indirectness in observer recept, information is presented through quantum phenomena; however the material paradigm veils paths for receptive relation understanding of information. Understanding information phenomena within the material paradigm leads not only to heavy fog in quantum theory but also generates more and more bizarre mysteries, It's not until a specific stage of advanced information research, closely related to information as receptive relation, that an important point with a fundamental nature becomes clearer at a deeper level: what we perceive with our eyes doesn't manifest something we don't have a concept of in our minds. The reason for this, as understood through receptive relation of information, is also clear: due to inertia within the material paradigm and the deep involvement of humans as receivers, understanding of information is inevitably shrouded by the material paradigm until the development of information technology brings about more and more fully manifestation of information.

4 The Veiling and Unveiling of Information

4.1 Introduction

Quantum phenomena have long presented themselves as manifestations of information as receptive relation, but we have long turned a blind eye to this fact, and might even, while concerned with the receptive relation understanding of information, we may have such doubts: "Why receptive relation?" Even secretly pondering: why bother? The reason for this lies closely connected to the following facts: humans are both biological beings and information agents. Due to the basic status of biological existence and the prolonged dominance of the physical aspect during evolution, just as the mind has long been oppressed by matter, humans have remained enshrouded in the fog of the relationship between matter and information.

In the understanding of information, the material veil is pervasive, much like a shadow following its figure. The reasons behind this are exceedingly complex, stemming not only from the deep-rooted nature of the material paradigm and its enduring inertia but also from the fact that humans, as the most highly developed information receivers known thus far, are deeply immersed in the understanding of information. This immersion is not only connected to the uniqueness of the information paradigm but also to the urgency of deepening our understanding of information.

Being the highest-developed information receivers to date, humans as receivers are deeply involved in the understanding of information. Therefore, we are often in the situation of not seeing the true face of "Mount Lu," only because one is in the mountain. Due to the long-standing inertia of the material paradigm, it is difficult to avoid a material understanding of information, leading to the veil of information by matter and energy.

In the study of the material paradigm of information, humans, as information receivers, find themselves in a situation similar to the classical physics concept of the human observer. In Newtonian mechanics, the observer takes their own standpoint as that of God. Viewing the world through the eyes of God, everything seems purely objective. In this sense, just as the theory of relativity and quantum mechanics brought about a Copernican revolution in physics based on the physical study, the study of information based on the material paradigm requires a similar shift. The paradigm shift involved here is even more fundamental than the paradigm shifts implied by the theory of relativity and quantum mechanics. The paradigm

DOI: 10.4324/9781003484851-4

involved in modern physics only concerns the realm of physics, while the shift of the information paradigm goes beyond matter/energy and involves information as receptive relation. The development of quantum physics has allowed this Copernican transformation process to unfold, but it is only with the emergence of information technology that we can clearly see this historical triumph. Information as receptive relation has been apparent in quantum physics for a long time, but the prolonged veil by the material paradigm has required the unveiling through the foundation of information technology. This illustrates the fundamental nature of the information paradigm shift. One of the typical examples of its importance is the manifestation of information through the development of information technology.

Regarding the importance of the information paradigm shift, this peculiar phenomenon in information understanding serves as a typical illustration. Quantum phenomena directly present information as receptive relation, but we have remained blind to this fact until the manifestation through information technology. Only through the manifestation of information with the development of information technology, which brings about the information paradigm shift, can we recognize that quantum phenomena are typical facts about information. Therefore, the manifestation of information through information technology also signifies the development required to deepen our understanding of information.

In many fields, our use of information seemed problem-free until the contemporary development of information technology, especially in the realms of big data and artificial intelligence, highlighted the profound issues in understanding information. On one hand, the connection between the core mechanisms of artificial general intelligence and the mysteries of consciousness and life increasingly points to the deepening of information understanding. On the other hand, the development of information understanding has led us to pay attention to dilemmas such as "information conservation."

The development of human civilization is a process of continuously unfolding information and a process where the need to deepen our understanding of information keeps developing. Including the development of industrialization, in the past, information understanding was fragmented across various disciplines, and to some extent, this fragmentation did not pose significant issues due to each domain excelling in its own way. With the development of information technology, on one hand, information itself gradually unfolds, leading to the increasingly apparent integration of various disciplines, necessitating the crossing of traditional disciplinary boundaries for information understanding. On the other hand, humans are rapidly transitioning from primarily existing in a physical way to increasingly existing in an informational way, making the need to deepen our understanding of information in human practice increasingly urgent. Currently, breakthroughs in the core mechanisms of artificial general intelligence and the associated quests to solve the mysteries of consciousness and life make it increasingly imperative to understand information more comprehensively. With big data and artificial intelligence as symbols, the contemporary development of information technology has fully manifested information and provided the conditions of our time for the removal of the veil of information, furthering its manifestation.

An information veil comes in two fundamental types: material veil and ideational veil, which primarily stem from the understanding of information as material and ideational or their attributes. The former arises when we understand pheromones or biological genes as information; within the material paradigm, this understanding inevitably leads to the veil of information. The latter mainly involves equating information with images, bits as data, or even knowledge. When the information paradigm has not fully matured, attempts to move beyond the material paradigm of information understanding often gravitate toward the ideational end, which is the opposite of matter and energy. Whether it is a physical or ideational veil, both are closely related to the unique situation of humans in the understanding of information.

4.2 The Unique Situation of Human Involvement in Information Understanding

Humans have not only been dealing with information, but their development is also a process where the role of information has become increasingly prominent. However, research on information started relatively late, to the point where it seems astonishing. "it is only in the last one hundred years or so that attempts have been made to create mathematically rigorous definitions for information" (Muller, 2007, p. 1). What's more, since its inception, systematic research on information has primarily manifested through the development of scientific information theory, and research on information philosophy, which is more of a philosophy of a science and technology nature, is a more recent thing. The fragmented nature of the discipline of information understanding is due to the categorization approach unique to the matter sciences. It was only with the contemporary development of information technology, particularly represented by big data and artificial intelligence that the conditions of our time were provided for the integration of scientific and philosophical research on information. This situation in information understanding has long overlooked a fundamental fact: the relationship between humans and information is entirely different from that between humans and material, making the human condition in information understanding highly unique.

In information understanding, the human condition is entirely distinct from the understanding of matter and energy. However, humans inevitably tend to analogize the study of information to the study of matter and energy, thus conducting research on information to varying degrees within the material paradigm. On the one hand, this is a normal progression in the development of research paradigms. On the other hand, due to the fundamental differences between the information paradigm and the material paradigm, it significantly limits the comprehensive progress of information research, resulting in a scenario where human understanding of information is entirely different from their understanding of matter and energy.

The relationship between humans and information has a developmental process that is fundamentally similar to the relationship between humans and matter/energy. However, it has a completely different historical origin. On one hand, just like the relationship between humans and matter, the relationship between humans and information primarily involves practical usage long before theoretical understanding. It initially revolves around the practical use of information, and then, with the gradual development of practice and understanding, it deepens its knowledge to make better use of information. On the other hand, the history of the relationship between humans and information can be roughly seen as equally long as the history of humanity itself. Compared to the history of the existence of matter and energy, the history of human existence is extremely brief, and compared to the history of the existence of information, human existence is relatively approximately identical. In relation to the developmental hierarchy of human information, the information of previous plants and animals can almost be disregarded – this itself touches on the critical difference between philosophical information theory and scientific information theory. The historical differences in the use and study of information by humans have determined their historical relationship with information. Based on the understanding of information as receptive relation, these two aspects of the relationship between humans and information allow us to better grasp the uniqueness of the human situation in information understanding.

In the relationship between humans and matter/energy, although humans are also physical beings, the fact that information as receptive relation can be ignored when considering humans' relationship with objects does not affect their understanding. The effectiveness of the dichotomous framework of subject and object in cognition has long been established in the historical relationship between humans and matter/energy. However, in the relationship between humans and information, humans, as information receivers, not only directly participate in constituting all information as receptive relation but also face an unprecedented paradigm gap because their relationship with information is distinct from their relationship with matter and energy.

The paradigm gap formed by the material paradigm and the information paradigm signifies the most fundamental paradigm shift. All the paradigms previously formed by humans belong to the material paradigm, with the information paradigm being fundamentally different. Therefore, in information research, when the paradigm shift has not yet been fully realized or implemented, various ways naturally attribute information to matter and energy. What is paradoxical, however, is that this fact itself can only be clearly seen in the understanding of information as receptive relation. What is presented here is the unique situation of humans in information understanding: upon entering the field of information, due to information feedback, a continuously complex bidirectional cycle mechanism is involved. The complexity of human understanding of information and the transition from the material paradigm to the information paradigm can be glimpsed through this.

From this, it can be seen that the relationship between humans and information is extremely unique, especially as self-consciousness beings, and increasingly as beings primarily defined by information. The development of information, as receptive relation, implies the development of receivers and sources of information. Within the process of information development, the integration of receivers and sources of information enables the emergence and evolution of information agents. When receivers can engage in complex information processing, they possess the ability of information production, leading to information evolution in an agent way.

As complex information agents with the capability to produce information, humans are not only the most advanced receivers but also the producers of information itself. Humans are not mere producers of material production; this fundamental distinction between information understanding and material understanding underscores the intricate situation of humans in the study of information. In the understanding of information, the most distinctive aspect of humanity lies in its deep involvement in the ongoing evolution of information.

In the comprehension of information, the deep involvement of humans makes their situation entirely distinct from their understanding of matter and energy. Undoubtedly, the study of information must go through a process similar to the study of matter and energy, progressing from complex composite forms to the most basic forms. Additionally, just as not knowing the fundamental structure of matter does not affect our everyday understanding and use of it, not knowing the nature of information not only does not hinder its usual use but also creates inconvenience when used at higher levels of abstraction (let alone at the basic level). However, precisely because of this, what is a natural scenario in the understanding of matter becomes a unique challenge in the comprehension of information.

From the perspective of receptive relation understanding of information, because information is a process of receptive relation between receivers and sources, humans, as the most advanced receivers to date, differ greatly from their understanding of matter and energy. Humans are deeply involved in understanding: understanding receptive relation between receivers and sources as receivers themselves. Consequently, their situation and what they recept are highly unique: all human recept, even their understanding and holistic guan zhao of concrete things are all based on the premise of the point of intentionality. This implies that wherever intentionality goes, it is the moment when receptive relation is established – where intentionality resides – even pointing to information existing, much like the sunlight illuminating a myriad of things. As the most developed receivers, when humans interact with sources using their sensors and establish receptive relation, their feelings are akin to receiving a certain kind of information, just as when they interact with physical objects, they feel like they are directly touching the "inherent" nature of those objects. This is precisely the typical phenomenon we observe in the presentation of information in quantum physics and, in fact, it is quite common.

Since it involves human observation, any matter science involves information as receptive relation. However, in descriptions of the existing physical world based on observation, receptive relation is often treated as properties of objects, and information has long been overlooked. This unique situation leads people to naturally feel that information is everywhere, and they even believe that there is nowhere without information. Just like Burgin put it, "Information is all around us, as well as in us" (Burgin, 2009, p. vi). What sensory receptivity points to are always concrete things existing as matter, and where intentionality points to, information exists. This naturally implies an understanding of information as something material. Even starting from the premise that information is not matter and energy; reducing information to properties of matter and energy still doesn't escape the "hands palm of the Tathagata Buddha"[1] in understanding information. For this reason, the situation of humans in understanding information determines that information, as receptive relation, is always in some kind of unopened state. As for what kind of information intentionality points to, it is closely related to the nature of the intention of the receiver; much like the color of an object seen by the human eye largely depends on the wavelength of the light projected onto the object and the specific receptive effects of a particular eye to certain wavelengths of light. This aligns with an ancient Chinese poem that resonates at the deepest level of human knowledge: "Not knowing the true face of Mount Lu, only because one is in the midst of it."

Due to its unique situation as the highest level of information accommodation, human understanding of information also finds itself in a very special situation. In the dichotomous paradigm of subject and object, everything that the human eye perceives belongs either to the nature of the object itself, such as the green color of tree leaves seen by a person being the actual color of the leaves, or it is attributed to the projection of the human eye, such as the eye perceiving that tree leaves as green. In either case, there is no existence of information as receptive relation. If one were to define information in this context, there would only be two possibilities: either it is seen as source-preexisting or receiver-assigning, with signal-carrying understanding being an extension of the source-preexisting understanding. It is this situation, that gives rise to the unique way in which humans understand information.

About how human understanding of information operates within this context, Jon Barwise has provided a vivid illustration.

Perhaps coming up with a theory of information and its processing is a bit like building a transcontinental railway. You can start in the east, trying to understand how agents can process anything, and head west. Or you can start in the west, with trying to understand what information is, and then head east. One hopes that these tracks will meet, but Fodor's paper "Information and Association," in this issue, tries to prove that they won't.

(Barwise, 1986)

Human cognition regarding information is carried out in this way. In the dichotomous framework of subject and object that characterizes the material paradigm, the effects of receptive interaction as information that result from interactions between subject and object are always attributed to the properties of the object and effectively "erased."

The deep involvement of humans in the understanding of information, as well as the differences between everyday and scientific uses of the concept of information, is significant. In everyday life and classical scientific fields, people can use the concept of information at different levels of understanding, depending on their specific needs and the context of daily life or a specific scientific discipline. These various concepts of information may differ significantly, but as long as they stay within the boundaries defined by specific paradigms, they generally work without major issues. This is why the process of human understanding of information is so uniquely important. What is even more intriguing is that this special "erasure" phenomenon in the understanding of information not only doesn't lead to a sense of loss but rather enhances convenience, particularly in practical applications.

For scientific information theory research, which operates within the material paradigm and is deeply rooted in it, this understanding of information is more of an advantage than a hindrance in the practical applications within specific scientific disciplines, much like the everyday use of the concept of information. However, this also means that in all fields and questions involving information as receptive relation, understanding and research cannot deepen or truly progress unless they acknowledge this "erasure" within the subject-object dichotomous framework. Therefore, recognizing this "erasure" within the subject-object framework is not only essential for philosophical investigations into information but also increasingly crucial for contemporary developments in science and technology, especially information technology.

This realization highlights not only the importance of transitioning from the material paradigm to the information paradigm but also the unprecedented level of difficulty associated with this paradigm shift. As an information agent that has evolved over a long period in the physical realm, and has always existed mainly in a material way, human has been deeply entrenched in the material paradigm. Consequently, the "erasure" of information to some extent within the subject-object framework is because the understanding of information is filtered through the colored lens of the material paradigm. Thus, in the human context of understanding information, not seeing the true nature of information is perfectly natural.

As human understanding delves deeper into the realm of information, there is another intriguing phenomenon: in almost all other fields, limitations in the level of understanding can lead to overlooking certain aspects of the objects of knowledge, but in the field of information, the human tendency to overlook information arises from our cognitive advantage.

In the understanding of information, recognizing the unique situation of humans as receivers deeply involved in the process leads to the "erasure" of information itself within the subject-object framework not only involves a deeper exploration of information but also necessitates a paradigm shift in both scientific and philosophical research, as well as their integration at a deeper level. The difficulty of such a paradigm shift is evident, but achieving it means that the human situation regarding information is no different from the situation regarding the physical world. By approaching information with the same orientating thought and mindset as we do within the physical world, we can truly arrive at philosophical thinking. It's not about reality being relational, or that relation is more fundamental than entity, it's that everything we perceive with our senses is not the naked essence of objects themselves but rather information as receptive relation which means establishing a relation. We understand the external world through these relations, and there is no escape, without exception. This situation, compared to projecting quantum theory onto Newtonian mechanics, is even more apparent and has more profound implications. Moreover, in a more fundamental sense, humans themselves are, in fact, relational beings, and the notion of transcending relation to reach the essence is a naive imagination. Furthermore, this is not a lamentable fact; it is, in fact, the glory of the evolution of biological intelligence into advanced stages.

From the perspective of human understanding of information, we can observe a series of peculiar phenomena in the history of human cognition. When people confronted the physical world, they embarked on epic, long-term exploration, unraveling a series of mysteries about the physical world; and when humans themselves increasingly exist in informational ways, they come to realize that the deeper enigma they face is information itself. In the history of information understanding, the concept of information has seen nearly 200 different definitions, and a general consensus on its definition has long been difficult to establish. This is due to the deep involvement of humans when facing information as the highest form of informational existence and its close relationship with the development of information technology. These two aspects together create a special situation in human understanding, where information is veiled.

4.3 Material and Ideational Veil in Information Understanding

In the face of the quantum phenomenon manifestation of information, one of the basic reasons why we remained oblivious for more than a century is fundamentally due to the material and ideational dual veil of information.

4.3.1 Material Veil of Information

The material veil of information in information understanding ultimately stems from the material paradigm, and its objective basis is closely related to the material occurrence of information and its subsequent inseparable relationship with matter and energy thereafter.

The mechanism of material veiling of information in information understanding is rooted in the human understanding of the material world through the material paradigm. To a large extent, understanding information through the material paradigm implies understanding information physically. In daily experience and classical physics, the material research paradigm typically manifests as a dichotomy between subject and object, and the cognitive structure of this dichotomy itself has the nature of a material paradigm. The reason the material paradigm is what it is lies in its disregard for information, and the subject-object dichotomy is premised on the assumption that information as receptive relation, which can be ignored as receptive relation, does not have a place within it. In this dichotomous framework of subject and object, there is no room for information. Since information, as an effect of receptive interaction can be said to be "erased" to some extent, only matter but not information is seen. As receptive relation, information is either attributed to the nature of the source or considered to be closely related to the attributes of the receiver. The obstruction of quantum phenomena presented by information in the material paradigm is a typical example of material obstruction of information.

In fact, due to understandable disciplinary reasons, understanding information physically is most typical in physics. The concept of the "Qbit ocean" proposed by Wen Xiaogang (Xiaogang, 2018) and related theories have had a significant impact on theoretical physics because they not only represent the highest achievement in understanding information and its relationship with matter and energy within the material paradigm but also reach a level of integration between science and philosophy. Through the concept of the "Qbit ocean," we can not only understand and explain quantum mysterious phenomena such as entanglement but also construct a more perfect picture of the physical world within the material paradigm. Although this represents a typical source-preexisting understanding of information, it is the most intelligent theoretical construct for understanding quantum mysterious phenomena within the material paradigm. Examining the concept of the "Qbit ocean" in reverse reveals that information is fundamentally veiled by matter due to the colored "filter" of the material paradigm. Thus, the material veil of information in human cognition is rooted in the material paradigm, and the formation of the material paradigm is based on material occurrence of information as an important objective foundation.

The material veil of information in information understanding is primarily rooted in the material occurrence of information. As information, based on physical receptivity, it initially manifests as physical activity. In the activities of organismic receptivity, information and matter/energy are in a relatively undifferentiated state since there is no differentiation between the source and the receiver of information. Therefore, the understanding of information through physical coding is almost inevitable. This is why in DNA, the coding of information through physicality remains intertwined with the matter/energy being coded, even though large molecules like proteins constitute the genes in DNA. Thus, the more primitive the stage of information evolution, the more difficult it is to distinguish between information and matter/energy. This is a typical extension of understanding information through physicality. Due to the fundamental similarity between the existence of information and matter/energy, the relationship between information and its encoding becomes obscure. Even in the process of differentiation between receivers and sources of information, when typical information has not yet emerged, information is more easily veiled by the physical sources of information in direct relationships between receivers and sources. This is a result of the unique dichotomous framework of subject-object understanding within the material paradigm, and it is almost inevitable that classical physics ignores information as receptive relation.

Due to the inherent dichotomous framework of the material paradigm, what appears before humanity in practical terms is a peculiar scenario in the understanding of information. On one hand, information is seen as a manifestation of physicality, and to grasp what information is, one must understand it within the realm of matter and energy. On the other hand, information seems to belong to the subjective domain, and there's no pressing need to delve into it for practical use. Almost instinctively, from a "God's eye" perspective, the objective existence of information appears indistinguishable from matter and energy. In this situation, within the dichotomous framework of subject-object, information is either understood as varying degrees of matter and energy or reduced to ideas related to various material objects, especially the new concepts emerging in the field of information. In essence, information itself is thereby "erased," and we are unable to perceive information as receptive relation, let alone comprehend that information fundamentally differs from matter and energy or ideas, and its real nature of relationality existence within concrete processes.

As a receiver integrated with the source of information, in the understanding of information, humans naturally tend to equate the source with information itself, as understanding information is essentially the production of information. For receivers capable of producing information, recept itself is the establishment of receptive relation, the creation of information. Therefore, within the specific dichotomous framework of subject-object inherent in the material paradigm, treating the source as information itself is quite natural. In such a scenario, viewing information as receptive relation becomes unnatural instead. Furthermore, for humans, who represent the highest and most complex level of receivers with information-producing capability, almost everything within recept is perceived as information itself. It is this deep involvement of humans in information understanding, driven by the material paradigm that results in multiple layers of material veil of information in understanding.

With the development of information material encoding, the material veil in information understanding has become increasingly diverse, even extending to the understanding of the material carriers of information. As long as information understanding is incomplete, there is bound to be a connection between information and its material encoding, resulting in varying degrees of confusion between information and the material encoding of information. Because information understanding in the material paradigm begins with matter and energy, and the material encoding of information is essentially an extension of matter and energy into information, information and the material encoding of information are almost inevitably confused in various ways. Within the veil of physicality in information understanding, the concept of an "information carrier" also contributes to the confusion. In receptive interactions based on matter/energy and having information natures, both matter/energy and symbols can serve as carriers of signals, but neither is the direct carrier of information itself. They are products of the materialization and ideationalization of receptive interaction effects and their processing – either unintentionally or purposefully produced information encoding or potential sources of information.

In the process of information generation and transmission, information coding like electrical signals is indispensable. Limited to a specific discipline scope will cause material obstruction in information understanding. From the point of view of information as receptive relation many concepts of "information carriers" are actually misuses of informational coding or signal carriers etc. Strictly speaking, there is no such thing as "information carrier." The so-called "information carriers" are products of signal-carrying, source-preexisting and receiver-assigning understandings of information. Moreover, as a conduit for signal transmission, the channel also plays a crucial role in the adhesion of information and signals, thus contributing to the material veil of information.

4.3.2 Ideational Veil in Information Understanding

Ideational veil in information understanding ultimately stems from the failure to achieve a transformation of the information paradigm, and this transformation involves deeper philosophical foundations. Closely related to the information paradigm, one of the most significant philosophical implications of information is the profound understanding of the world in terms of relationality, based on the understanding of the world as material entities. This understanding becomes clearer after achieving the transformation of the information paradigm.

In the material paradigm, without the relational items, relations always seem ethereal, and the items of the relation are considered more fundamental than the relation itself. It is for this reason that, in information understanding, it appears that only sources, receivers, and information encoding are the actual existents. Understanding information as receptive relation not only highlights the relationship between information and information encoding but also positions information as a pivot jointly constructed by both sides in the cognitive relationship between humans and the world – this is the Archimedean point that humanity can truly find. The clarification of the relationship between information and information encoding deeply involves the most typical manifestation of information being veiled by matter and idea. Information, as the Archimedean point of human cognition, becomes the pivot of the transformation of the information paradigm.

In the complex relationship between information and information encoding, the most veiled aspect involves bit, which has increasingly been widely regarded as information itself. As a digital encoding for information, bits can be directly associated with material encoding, such as the "on" and "off" of electrical currents. On the other hand, they are neither matter nor energy; they are, in fact, digital encodings in information symbol encoding, and the "0" and "1" that constitute information encoding are ideational encodings of information. It is through this extremely veiled way that the most typical ideational veil in information understanding is formed: the digital encoding veil of information, with bits widely regarded as information itself.

Ideational veil in information understanding takes many forms, from simple information encoding to complex ideational systems, all of which are understood as information itself, constituting a complex ideational veil of information. The most common is understanding information as data, which is a generalization of the bit understanding of information. At its extreme, some even equate knowledge with information, believing that information is knowledge.

An ideational veil of information primarily manifests when, on the one hand, humans have developed complex concepts and conceptual systems based on information encoding, and on the other hand, due to the lack of research on ideational coding of information, it is impossible to understand the relationship between concepts, knowledge, and information. Therefore, as the highest and most complex level of agent, the complex information, information encoding, and conceptual systems of humans constitute a veiling of the basic form of information. Among them, the mechanism of complex information veiling is not only the result of information evolution but also requires a transformation of the information paradigm through higher-level information evolution to bring about a clear change.

It is the development of information technology and information civilization that allows us to see that, in the ultimate sense, biological evolution is information evolution in essence, and humans are the highest-level agents in the current stage of information evolution – agents with self-consciousness. Developed as an outcome of the integration of receivers and sources, human agents stand far above the basic forms of information, and the interactions are primarily with complex information composite forms, so it is natural to lose sight of information itself. For the same reasons, in practical use at the material level, it is not only unnecessary to precisely depict the basic form of information but also more natural and reasonable to consider information as a concept in the material paradigm. This can only be seen after the transformation of the information paradigm, and this is the root of ideational understanding of information, as well as the paradigmatic reason for the ideational veil of information.

4.3.3 Paradigmatic Dilemma of Information Veil and Its Significance

Whether the material or ideational veil in information understanding, as the development of information itself progresses, the paradigmatic dilemma and its significance in understanding information become more and more apparent.

In the context where information can ultimately be reduced to matter and energy or their attributes, the material veil of information implies that information is naturally treated as pure objective existence just like matter and energy in the material paradigm. However, as information understanding deepens, the objective interpretation of information becomes more apparent, leading to a growing paradoxical situation of paradigm. One of the most typical examples of this is when divorced from specific context, the assertion "information is physical" can imply that information is equivalent to physical attributes. This can create a significant barrier in information understanding, as it constrains information in different ways from physical attributes within information comprehension. The most extreme way in which the material veil of information highlights the paradigm dilemma is that it ultimately leads to an entity understanding of information.(Landauer, 1999) The entity understanding of information is precisely the most fundamental aspect of highlighting the paradigm dilemma in understanding information: almost all important characteristics of information that are different from that of matter and energy cannot be understood by the tendency to understand information as entity.

In the realm of physics, the entity understanding of information evidently treats information as its material encoding, even venturing into the "physical properties of information," effectively equating to the primary natural source or the information encoding as a secondary source – often represented as information products.

Paradoxically, information products can have entirely opposing meanings in the context of information comprehension. One meaning is undoubtedly the material understanding of information. The other, however, may lead in the right direction for comprehending information, depending on whether it's within the matter or information paradigm. In human understanding, dilemmas are often associated with breakthroughs at higher levels. This scenario is most typical in the context of information understanding.

Within the material paradigm, even if one were to push it to the extreme by reducing information to the "representation of events" and their reflections, it might still be challenging to truly grasp information because representations can be subjective reflections of objective entities. This is where the existential root of the human tendency to mistake the source for information itself lies. In this situation, individuals who serve as specific receivers of information may unconsciously perceive this secondary source as information itself, and not only do they sometimes consider this potential secondary source as information, but they might also go a step further and view information as an entity independent of the receiver. It can be observed that as the most complex form of information so far, human understanding of information is often influenced by their deep engagement. As potential receivers with highly developed receptivity, when humans intentionally illuminate potential sources, establishing receptive relation, they can easily mistake the potential source for information itself. As a result, they might believe that information (or at least the source) exists objectively even in the absence of the receiver, just as matter and energy exist in the natural world even when not perceived by humans.

In the information paradigm, however, the results can be entirely different. Viewing information as the "representation of events" may actually facilitate a closer understanding of information because representations are inherently closely tied to receptivity. Information may thus be revealed as the process of receptive interactions as relation. The establishment of receptive relation between the receiver and the source, based on the characteristics of the receiver, in the way available for certain receivers rebuilding specific receptive relation, produce information encoding – examples include punched cards, patterns on records, and tracks on tapes. Since particular receivers can reconstruct specific information as receptive relation on these examples, these information encodings become information resources as potential secondary sources. As products of material information encoding processing, information products with physicality possess objectivity. When people realize this objective existence, they can approach a more reasonable understanding of information, albeit usually only achievable through the development of information technology.

Moving from information products to information is a reliable approach to grasping information, and it primarily points toward information technology. Specifically, the development of digital data is critical in the unfolding of information within information technology. The unprecedented unveiling of information in the development of information technology is widespread, with the development of big data and artificial intelligence being the most typical examples.

4.4 The Big Data Unveiling of Information

The material and ideational veil of information has persisted throughout the entire process of information understanding due to the deep involvement of humans in this process. The systematic study of information by humans initially began in the field of communication science, where the nature of signal processing determined that grasping information primarily involves quantification. Since mathematical formalization is regarded as a crucial criterion for science, Shannon's information theory is widely recognized as representing the most fundamental definition of information. As Scott J. Muller discovered, "it is only in the last one hundred years or so that attempts have been made to create mathematically rigorous definitions for information" (Muller, 2007, p. 1). However, the mathematical definition of information is not synonymous with a general definition of information, which must be grounded in the integrated development of science and philosophy.

Since information itself is an existence in integration of science and philosophy, the development of human information technology not only unfolds the nature of information itself but also increasingly showcases the integrated nature of information in both scientific and philosophical contexts. In the unveiling of information through information technology, the more advanced the development of information technology, the more comprehensive the unfolding of information becomes. This is why it is only in contemporary times, with the advent of big data and the subsequent development of artificial intelligence, that we find the conditions to expand our understanding of information to a specific extent.

The unveiling of information through information technology implies the removal of the material and ideational veil of information. Big data, as a product of the development of digital encoding, has established a world of informational relation – a world that possesses a platform-like nature for humans, equivalent to the material world, and even richer in terms of relation and possibilities. Big data, as a collection of digital encoding of information, does not exist as a material entity. For both its nature and its significance to humans, it is fundamentally a realm of relation – a world of correlations. Big data, as the result of digital encoding development, is precisely the unfolding of the relational aspect of information. On one hand, this provides concrete insights into receptive relation understanding of information, and on the other hand, it advances the deepening of big data understanding of information. The relational existence formed as a result can only be fully understood and utilized based on the receptive relation understanding of information. These two aspects constitute a mechanism of a bidirectional cycle between big data and information understanding.

As an informational existence, big data exists not only fundamentally different from the natural world but also intimately connected to human existence. Unlike the natural world, which stands independent of information as receptive relation thereby human being, big data is purposefully collected by various sensors and generated through the digital encoding of receptive relation. It is precisely as a product developed by digital coding of receptive relation that big data and its contained correlations not only make information as receptive relation but also make unprecedented unveilings in the evolution of information. In essence, the unveiling of information through big data is rooted in the development of digital data. Unlike analog data, which is often inseparable from its material encoding, digital data is explicitly detached from physical, indicating that it is not a form of material existence. Therefore, as a product of digital data development, big data does not exist in a material way. It is distinct not only from natural objects but also the receptive product of digital technology and equipment. The existence and development of big data constitute a process of unveiling information.

Because big data is a product of the development of digital information encoding, the unveiling of information through big data is fundamentally the unveiling of digital information encoding. However, in the unfolding of big data development, digital encoding alone cannot fully reveal information. Data is a form of information encoding, and information encoding has undergone a transition from material, analog encoding to ideational, digital encoding. Therefore, general data has existed almost since humans used information. However earlier data was more direct material coding; more direct human activities were carried out with objects such as "fingers" in "counting fingers" and "ropes" in knotting records; as it developed later on, data had less pure object-coding properties such as from counting stones to abacus beads which although were material coding in objects had different calculation mechanisms. However, material coding like abacus beads and ideational coding like natural numbers had completely transitioned from material analog coding to ideational digital coding in object coding. Since pure ideational coding like natural numbers cannot establish direct interaction mechanisms with physics they can only be carried out through humans thus they cannot directly connect with empirical material mechanisms. Only when it develops digital coding can it establish direct connections between ideational coding and material coding in object at mechanism levels.

Because digital information encoding not only reaches the level of ideational coding but also establishes a direct connection with material information encoding, the unveiling of information through information technology is, essentially, the data unveiling of information, while the data unveiling of information is more the digital data unveiling of information; its mechanism directly connects processing for informational signals between digital data and corresponding material encoding. It is precisely based on digital data that conditions are provided to relatively separate the material understandings of information.

> We are asking whether the universe in itself could essentially be made of information, with natural processes, including causation, as special cases of information dynamics (e.g. information flow and algorithmic, distributed computation and forms of emergent computation). Depending on how one approaches the concept of information, it might be necessary to refine the problem in terms of digital data or other informational notions.
>
> (Floridi, 2011, p. 44)

Floridi's philosophical reflections on information underscore the significance of digital data in understanding information. Due to the direct connection between digital data and material information encoding, examining information through digital data means looking at information in its simplest and clearest form. It is akin to observing the fundamental nature of receptive relation through the most straightforward sensor, such as a camera.

Therefore, understanding data is crucial for comprehending information, but this importance is fully accentuated only after the digital data constitutes big data. The development of big data has made it increasingly evident that all these digitized products actually are data. All data is a product of encoding receptive relation established by various sensors. Data itself is not information, but it is treated as information and carries information value precisely because it is an information encoding or even an information product. It is here that the development of big data begins to unveil the mystery of information. In this process, the bidirectional cycle between big data and information establishes a mechanistic foundation for continually deepening our understanding of the information world.

The unveiling of information through big data sheds a gradually brightening light on the challenging process of understanding information. As a backdrop comparable to a holistic theory of abstract universality, big data is an informational existence produced by humans, available for sharing among all beings capable of information production. As a product produced by human information, big data constitutes an integrated world of information. Facing a single source, humans as receivers easily regard it as information itself; while big data is formed by humans with abilities to produce information for their own production, it is precisely its difference from material natural sources that allows matter and information to be relatively separated. It is precisely its integrated existence in the world of big data, with its nature different from the material natural world, that makes information present as receptive relation, thus allowing information to be revealed: information is not a purely objective existence unrelated to the receiver and its receptivity as the most basic unit, information can only exist as receptive relation. The process of information production by the receiver is a process of receptive relation, and the objects and products for this kind of production are (mostly) not information itself but rather various information resources as information encoding.

Before the development of digital data, especially big data, realizing that "information indeed exists relatively, both source (object) and receiver (subject) are indispensable" was already not easy; however from this perspective, "Information is the indirect existence of the characteristics it represents" (Kun, 2017, p. 467). Although the transformation of the information paradigm has not yet been completed, it is already the highest achievement possible under such historical conditions. Only in the digital age can we see a deeper direction in understanding the mechanisms of sender and receiver relationships. In the material veil of information, finding the correct path can present unexpected difficulties, stemming from historical limitations. Before the digital age, it was impossible to find the right direction, but starting from digital data, it has become almost the natural path for information thinking.

From this, it is evident that for humans, who are advanced receivers of information, the signal-carrying understanding of information is quite natural. Similarly, the source-preexisting understanding of information is also natural and inherent. In fact, this understanding predates the field of communication science and has existed in everyday life for a long time, but it was the development of digital encoding that provided important insights into the receptive relation nature of information.

The unveiling of information through big data is a concentrated reflection of the information unfolding of the development of information technology. Furthermore, the continuous expansion of information deepens our understanding of big data, leading to further unveilings of information. It is precisely in the big data unveiling of information that our understanding of information can go deeper into more basic levels.

The development of big data allows us to pose new questions about information. Revealing the relationship between signals and information almost enters into receptive relation understanding of information because having information and containing information is related to differentiating between potential sources and information encoding. In information understanding, the result of distinguishing between information and signals is that information is stripped from signals, and this kind of stripping product can only be a relational existence. This leads to the receptive relation understanding of information, which involves not only an awareness of how humans transform their own ways of existence but also an understanding that information resources are distinct from material resources.

Research on information encoding is related to the recognition of information resources. This not only directly relates to the way humans exist and develop as information beings but also affects how we can fully utilize and develop material resources for increasingly effective use, forming a bidirectional cycle of information production and comprehension through the unveiling of big data. And to associate information resources with the nature of human information needs not only has more profound meanings for human development but also deepening understandings of information.

A more accurate understanding of information not only involves the development of information technology but also involves the understanding of human needs and human development. Information understanding involves the understanding of human needs; just by understanding that information resources are entirely different from material resources, one can clearly see the importance of a more accurate understanding of information for comprehending human information needs. Confusing information with, or even equating it to, material resources lowers information needs to a level similar to material needs, obscuring the fundamental nature of information needs as distinct from material needs.

The significance of information needs for humanity, due to the inherent differences between material and informational natures, cannot be explained by material needs alone. Hence, to further comprehend, explain, and promote human societal development, the need to unveil information and move away from the material aspect becomes increasingly imperative. The unveiling of information is not only the unveiling of human characteristics but also an indispensable mechanism for individuals to pursue comprehensive personal development within the human species. Human development itself is a process of information unfolding. The development of the material forms used by receivers for information activities is precisely the development of the unveiling of information from its material aspect. This process is also the development of information technology. It is the development of information technology that makes it possible to lift the veil of the concept of information, and the unveiling of information through big data is a typical example of removing the obscuration of the concept.

The unveiling of information through big data is based on the development of data itself. In this sense, the former is the result of the latter's development. The concept of "data" is very broad, but in the context of "big data," "data" specifically refers to digital data because only digital data can constitute "big data" in the truest sense. Therefore, in the context of the mechanism of information unveiling, the unveiling of information through big data is synonymous with the unveiling of data in information terms. As for the results of this unveiling, the unveiling of big data is synonymous with the unveiling of information through big data. It is the development from data to big data that has brought about revolutionary changes in understanding information, as well as in the development of human civilization.

Big data implies a data revolution, which is an essential foundational aspect of the information revolution. Without the data revolution, completing the information revolution would not be possible, let alone reaching its climax. It is big data that has opened the door to human information civilization. The development of information civilization, in turn, deepens human involvement in understanding information. Hence, the unveiling of information through big data not only holds a deeper significance than the unveiling of digital data but is also of greater importance for human development.

The development of information technology signifies the unfolding of information, and this is prominently exemplified by the era of information civilization launched by the development of big data. What the unveiling of big data in information refers to is precisely this: the digital data world constituted by big data, which not only highlights the increasing importance of clarifying the concept of information but also emphasizes the position and nature of information as a basic constituent level.

The development of big data not only reveals the transition from digitization of matter to informatization of matter and from materialization of data to materialization of information in a profound sense but also leads to a bidirectional cycle structure constituted by these two processes. Consequently, it offers a higher-level holistic guan zhao for understanding information and its evolution. It is only between informatization of matter and energy and the materialization of information built upon big data, that we can achieve a genuinely deep understanding of information at the mechanistic level.

The development of information technology contributes not only to a direct enhancement of information understanding but also serves as an internal driving force for increasingly robust information research. The fundamental driving force behind information research resulting from the development of information technology is its demand for a deeper understanding of information. The contemporary development of big data, especially when combined with artificial intelligence, has unprecedentedly reinforced the need for more precise information research.

4.5 The Artificial Intelligence Unveiling of Information

Whether it is related to the breakthrough in the research of the core mechanism of artificial general intelligence, or the unraveling of the mysteries of life and consciousness, the key role of deepening information understanding is self-evident. If it were not for the research on artificial general intelligence, humans might not have been so eager to understand the basic form of information. As long as the focus remains primarily on the existence of physical entities, there may be less necessity to delve into the basic forms of information concerning human information activities.

The development of artificial intelligence has transitioned from being driven by human knowledge (typified by expert systems) to being data-driven (with recent examples like ChatGPT), and its latest advances come in the form of a large language model (LLM) driven by big data. With the growing ubiquity of LLMs, there is an increasing expectation that artificial general intelligence is on the horizon. Some experts even argue that these models represent the gateway to achieving artificial general intelligence. However, as statistical models, large language models gain their ability to converse with humans by bringing vast amounts of data into human contexts, but they lack genuine understanding themselves. In this sense, large language models cannot directly evolve into artificial general intelligence.

Despite their incredible search and computing capabilities, large language models far surpass individual humans in some aspects of language proficiency, but their understanding capabilities are nowhere near comparable to those of human children. This highlights the critical role of information as receptive relation in the development of human-like artificial intelligence. The evolution of artificial intelligence must progress from being data-driven to being information-driven, thereby not only further revealing information on a deeper level based on big data but also revealing its general progress against the core mechanism associated with the mysteries of consciousness and life, highlighting an increasingly urgent need: artificial intelligence, on the level of data as information encoding, further delves into the level of information as receptive relation. The development of artificial intelligence must go from human knowledge to the basis of big data, and then further deepen into information as receptive relation. It is in this sense that big data-driven artificial intelligence unprecedentedly highlights the importance and urgency of deepening information understanding. Only in a living scene and consciousness and life-like human intelligence can we truly understand the process of receptive relation as information.

The breakthroughs in the study of the core mechanisms of artificial general intelligence, connected to the unraveling of the mysteries of consciousness and life, not only demand an unprecedented level of understanding of information but also emphasize the importance of the development of receptive relation. Whether these breakthroughs must occur at the quantum level or not, can be seen through the holistic guan zhao of receptive relation on quantum phenomena that reveals a deeper level of veil of information understanding. This allows us to revisit the long-obscured quantum phenomena presentation of information, exposing the process and scenarios of its unveiling.

The material and ideational unveiling of information are typical scenarios within the information guan zhao on quantum phenomena. The receptive relation understanding of information reveals that, in quantum observations, due to the inability of the senses to directly build up receptive relation with quantum objects and the necessity for experimental arrangements, humans, as advanced observers (incompatible within the same macroscopic scale), avoid direct deep involvement. Thus, the perspective shifts from "where the intention goes is all information" to a manner resembling "backing away," enabling us to see scenarios that were previously challenging to observe. Consequently, in the quantum domain, an unprecedented landscape emerges – one that blurs the landscape of physicality while rendering the information landscape increasingly clear. Looking back at the macro field, quantum phenomenon, as an observational effect, is a typical receptive relation. In this light, when we reflect on the macroscopic domain, the situation remains the same. The only difference lies in whether the observed object and human sensory characteristics match. From this, we can not only see the information technology unveiling of information but also see its typical case: the unveiling of quantum phenomena presentation of information.

The crucial significance of quantum mechanics in relation to information lies in the separation of the observer and quantum objects during quantum observations, thus highlighting the receptive relation between the observer and quantum objects. This manifests quantum phenomena – the observational effect as receptive relation. Related to this, Timpson once raised a critical question:

> There is an important question which my deliberations throughout the course of this study have touched on repeatedly, but have not answered. This is the question of what the role of a concept like information is, or might be, in physics. In particular, the question of whether we ought now to recognize information to be a fundamental physical concept, as, for example, energy (fairly uncontroversially) is.
>
> (Timpson, 2013, p. 207)

The issue of the status of information in physics is closely tied to both the understanding of information and the relationship between quantum physics and classical physics. The status of information in physics cannot be clarified in terms of the signal-carrying, the source-preexisting, and the receiver-assigning understanding of information. Only through the receptive relation understanding of information can this question be reasonably answered in the context of the relationship between quantum physics and classical physics.

Because of its seemingly inexplicable nature within the framework of physical principles, quantum physics becomes a typical domain for the source-preexisting understanding of information. Regarding the tendency of the source-preexisting understanding of information, it essentially posits that information originates from the source of information, just as water flows from a water source. Hence, the source of information is the originating of information, and in the production of information, it is the generating end. This is why a signal, while dependent on the conditions of the source and channel, carries only information about the source. Among the conditions on which the signal depends, only the source of information is the producer of new information. Believing that information comes from sources, information as preexisting object flow from source to signals means that to understand information as an entity-like existence, sources are producers of (new) information which means channels to information just like a river way is to river water. Signals are just carriers for information; channels are existing conditions that do not generate (new) information. To consider that information originates from the source of information is to view the source as the creator of information, akin to viewing a relational term as the creator or basis of the relation, and this is in the sense of the material paradigm (entity orientating thought). What the material paradigm implies is a way of existence entirely distinct from information.

Even if breakthroughs in the core mechanisms of artificial general intelligence do not necessarily occur at the quantum level, they must take place at the level of the receptive relation understanding within quantum phenomena. This is the most profound unveiling of the generalization of artificial intelligence in the context of information understanding. If we compare artificial general intelligence to humans as humanoid intelligences, we can arrive at an explanation that goes beyond mere analogy: a deceased organism still contains its biological encoding of information, but it is a corpse; a comatose individual retains life but only possesses a process of organismic receptive relation, whereas a fully alive, normal human possesses both organismic and sensory receptivity processes. This illustrates the most concrete unveiling of the development of artificial intelligence in information understanding.

Artificial intelligence is the objectification of human self-awareness as a whole. By looking back at humans who already have mechanisms for general intelligence from generalizing development of artificial intelligence, we have obtained an indispensable mirror for more fully understanding ourselves. Humans and even animals already have core mechanisms of general intelligence, but we ourselves do not understand the mysteries of consciousness or even life. Linking it to the development of artificial general intelligence at the core mechanism level, the unveiling of artificial intelligence of information is even related to self-understanding of the way of human existence.

Compared with matter and energy that we have always been very familiar with, the existence of information is indeed quite subtle. In both scientific fields and daily life, people typically focus on the activities of signals. What information truly is, how much we know about it, to what extent we understand it, or even whether our understanding is accurate, often doesn't have a tangible impact within specific contexts. A prime example is the field of communication, where the research subject isn't information itself but rather signals. Communication science is the study of signal transmission and processing, not a science that delves into the most fundamental aspects of information. It's only in more foundational areas science and technology integrated that the precise definition of information becomes necessary. Currently, one of the most prominent fields in this regard is artificial intelligence. The development of artificial intelligence, especially its generalization, urgently requires a deeper understanding that transcends data as information encoding and delves into information as receptive relation.

Standing at the highest level of information technology development, we can see the panorama of information unveiling through artificial intelligence. The development of information technology not only manifests in the overall development of information civilization but also in the unprecedented societal demand for information. Recent research shows that the focus on digital encoding of information not only highlights the intricate relationship between information and matter/energy but also deepens our understanding of the relationship between source and receiver. The "information lifecycle" involves cycles in relationships between the source and receiver. Artificial intelligence, as the main emblem, signifies the unfolding of information technology and offers the conditions of this era for information research, primarily in the context of the paradigm effect of information unveiling.

The universal unveiling of information as receptive relation implies a continual expansion of the effects of paradigm shifts in information understanding. Human deep involvement in information understanding influences both our human understanding of information and, importantly, underscores our unique advantage in understanding information after its unveiling. In the realm of information understanding, human deep involvement transitions from a disadvantage to an advantage, hinging on the shift from the material paradigm to the information paradigm. (Just as in the material paradigm, information encoding veils information, while in the information paradigm, we can further unveil information through information encoding.) This establishes a bidirectional cycle between information understanding and the development of big data and artificial intelligence.

The development of information technology, especially in big data and artificial intelligence, not only presents the task of understanding information at a deeper level but also provides the historical conditions for accomplishing this task. A deeper understanding and grasp of big data relies on the deepening of information understanding, while breakthroughs in the core mechanisms of artificial intelligence must be grounded in a deeper understanding of information. If we consider the ultimate completion of the information revolution as achieving a thorough understanding of information, then this comprehensive understanding of information will undoubtedly lead to revolutionary transformations in the understanding and application of big data and the development of artificial intelligence. In the holistic guan zhao of information, big data, and artificial intelligence within the context of information civilization resemble a great tree of human civilization. Information forms the roots of this tree, serving as its foundation and essence. The roots of the tree of information civilization are deeply embedded in the soil of matter and energy. The soil of matter and energy is the foundation upon which the tree of information civilization stands, rooted in the earth of matter and energy. As we human beings are both the typical and highest-level receivers of information, the environment we find ourselves in is akin to "information Mount Lu." The fog shrouding our field of information research stems from being deeply situated within it. As the development of big data and artificial intelligence allows information to be unveiled, the emergence of information in big data and artificial intelligence is similar to the germination and growth of information as a seed, gradually yielding the foundation of big data for information civilization and the fruits of artificial intelligence.

4.6 Conclusions

In the realm of quantum phenomena, the presence of information as receptive relation has long been evident, yet we have often turned a blind eye to it. This intriguing phenomenon is closely linked to the unique situation of humans in the understanding of information. As a complex receiver, humans are deeply entwined in the comprehension of information. Due to the inertia of the material paradigm, our understanding of information is shrouded by a dual veil of materiality and conceptions. Within the material paradigm, the receptive effects generated by observation are attributed to the properties of objects, creating a barrier between the observer and the object, obscuring information as receptive relation in an erasing way. It is impossible for humans to avoid the material obscuring within information understanding with the inertia of the material paradigm, and the material obscuring in the understanding of information rebounds, giving birth to understandings of information such as data, knowledge and even images, constituting the ideational obscuring in the understanding of information.

The development of information technology, especially big data and artificial intelligence, has gradually unveiled information and promoted the shift from the material paradigm to the information paradigm, making information as receptive relation increasingly clear. The development of big data has made it increasingly evident that all digital products are data, and all data are products of encoding of receptive relation established by various sensors. It is through this development that the mysteries of information have started to unravel. In this process, the bidirectional cycle between big data and information lays a mechanistic foundation for deepening our understanding of information. The development of artificial intelligence advances the big data unveiling of information to a deeper mechanistic level. The development of artificial intelligence from human knowledge-driven to big data-driven levels brings about large language models represented by ChatGPT, whose generalization ability makes the general of artificial intelligence explode; on the other hand, it is destined not to be able to move toward artificial general intelligence because it does not have real understanding ability.

The big data and artificial intelligence unveiling of information means that there is an unveiling of information in information understanding. Looking back from this perspective, not only is the quantum phenomena presentation of information clearly visible, but also further presents a deeper connection between the understanding of the nature of source at the quantum level and the core mechanisms of artificial general intelligence.

The big data unveiling of information has deepened our understanding of big data and research on artificial intelligence with the developments of information civilization opened up by big data, systematic explorations about information are becoming more and more urgent for humans. The explosive demand for information driven not only by societal development but also by the increasingly enhanced information research needs of information technology, represented by artificial intelligence, constitutes a powerful driving force capable of pushing information research to its deepest levels. As products of perceptual sensors, data and the deepening understanding of their relationship with information become increasingly important, especially digital data. Data is the most typical informational way of existence that is completely different from material existence in the natural world, and the information world is increasingly composed of digital data. Hence, in the contemporary context, data has become synonymous with digital data, leading to the comprehensive unveiling of information in the digital age.

In the digital age, on the one hand, the situation of humans deeply involved in the field of information has changed unprecedentedly, and on the other hand, information continues to unveil itself in depth. The development of information technology has given rise to artificial intelligent agents beyond biological intelligence. The development of these artificial intelligent agents is unfolding the primal forms and mechanisms of agents, allowing humanity, as advanced agents and obsessing themselves, to gain insight into these agents. The unveiling of information as receptive relation implies the natural unfolding of information. In this sense, the unveiling of information ultimately manifests as the self-unfolding of information, which is the fundamental process of removing information veils.

The material veil of information by materiality has rusted the principle that information is neither matter nor energy in the understanding of information; while the ideational obscuring of information excludes many information activities outside humans from information. Equating information with data obscures at least the process of plant information; while equating information with knowledge excludes at least animal information activities. Whether pheromones and biological genes are understood as information itself or information is understood as data, knowledge, or imagery, it essentially involves the information encoding understanding of information. Therefore, the receptive relation understanding of information naturally raises questions about information encoding and its relationship with information, which is closely related to the complex relationship between information and its encoding. It is only through the information technology unveiling of information as receptive relation that the veils hiding information encoding can be truly removed. Therefore, to eliminate the material and ideational veils of information, there is a fundamental need for in-depth, systematic research on information encoding and its relationship with information.

In the understanding of information, the prolonged material overshadowing of information has understandable objective reasons, mainly due to the deep involvement of humans as information receivers. The prolonged failure to recognize the quantum manifestation of information can be attributed to the fact that information research has long been entrenched within the material paradigm. To completely remove the material and ideational veils in information research, a fundamental shift to the information paradigm must be realized. In the process of transitioning from the material paradigm to the information paradigm, systematic, in-depth research on information encoding and clarification of its relationship with information plays a fundamental role.

Note

1 A metaphor for being unable to get rid of the control of others or a certain force.

5 Information and Information Encoding

5.1 Introduction

The receptive relation understanding of information provides a premise for clarifying the relationship between information and information encoding, thereby laying the foundation for a more accurate understanding of information encoding. Based on the receptive relation understanding of information, information encoding and its relationship with information have become an important field of information research, with both existing questions and paving the way for new domains.

In the field of information research, our understanding of information actually begins with information encoding. On one hand, we often have a relatively better grasp of certain information encodings while remaining less informed about the information itself. On the other hand, the adequacy of our understanding of information encoding depends on our comprehension of information. Due to limitations in information understanding itself, the study of information encoding is, in reality, greatly constrained. Without a thorough understanding of information itself, it becomes impossible to grasp the relationship between information and its encoding, thus information encoding.

Information encoding is commonly understood as the process of imparting codes to information. In the field of communication science, this understanding primarily revolves around the transformation of information encoding between codes. "The word *code* usually means a system for transferring information among people and machines. In other words, a code lets you communicate" (Charles, 2000, p. 5). In communication, the purpose of information encoding is to ensure the rapid and accurate transmission of signals to the receiver, with source encoding aiming to convey the signal using as few symbols as possible, enabling the precise and swift rebuilding of specific receptive relation by the receiver. The rebuilding of receptive relation is connected to information decoding, while "channel coding" aims to minimize interference during signal transmission.

Due to its origins in communication science, the understanding of information encoding is closely tied to the nature of communication science.

Encoding is the process by which information from the source is converted into signals (specifically, symbols in the Shannon-Weaver case) to be transmitted. It is a set of exact and systematic regularities that can be built into

DOI: 10.4324/9781003484851-5

the message constructed by the transmitter. Decoding is the inverse process where the receiver converts the signals received signals back into understandable information.

(Fresco & Wolf, 2016, pp. 81–82)

This is entirely an understanding of communication science's view on information encoding, which does not cover all ways of coding or even physical encoding outside signals, let alone biological encoding.

In the ongoing process of information understanding, confusion arises between information and information encoding to varying degrees, leading to the blurring or entanglement of concepts such as information, news, facts, data, and knowledge in the development of information technology (Watson, 1976, p. 553). Some understandings of information encoding are linked to higher-level products and problem-solving solutions based on information encoding.

Information encoding is sometimes perceived as similar to programming, algorithms, or even software design (David & Lyn, 2022, p. 174). This obviously represents an over-layered understanding of coding using more complex methods that can explain things more clearly. Some views regard coding as a manner (Khosrow-Pour, 2007, p. 91). Thus it can be seen here that when there's no complete understanding of what information really means yet, coding tends to be understood in some specific way that doesn't cover all aspects or it's interpreted as something more complex at higher levels or processes. This trend became more and more obvious with the development of information technology.

In the field of information technology, due to the understanding of information, the development of digital information encoding has made the understanding of code conversion in information encoding increasingly important, and there is even a tendency to understand information encoding as the conversion between different codes. In the authoritative *Dictionary of Information Science and Technology*, the following content is listed as the primary definition of "encoding": "The process by which the content and meaning that is to be communicated is transformed into a physical form suitable for communication" (Khosrow-Pour, 2007, pp. 228–229). It can be seen that the understanding of information encoding is inherently connected to our understanding of information. The premise of information encoding understanding is the understanding of information. Only when we have a solid grasp of information can we understand information encoding properly. Otherwise, we may mistakenly interpret information encoding as a broad concept of encoding or even misunderstand general coding as information encoding. A typical example is viewing the conversion of text into binary digital code as information encoding.

The understanding of information encoding and information itself is a sequential development process, with information understanding serving as a prerequisite for information encoding understanding. Without a solid understanding of information, it's impossible to have a comprehensive understanding of information encoding. Based on the receptive relation understanding of information, a more accurate and systematic understanding of information encoding can be achieved.

The receptive relation understanding of information suggests that coding can be generally categorized and narrowly defined. In the general sense, coding implies the conversion of representation, leading some to believe that encoding is simply the conversion between different codes; while narrow coding specifically refers to information encoding. Because representation does not require an explicit decoding process, as a conversion between different representations, general coding is just a conversion between representations. It needs decoding only when the converted representation aims at allowing specific receivers to rebuild corresponding information as receptive relation "The word *code* usually means a system for transferring information among people and machines. In other words, a code lets you communicate. Sometimes we think of codes as secret. But most codes are not. Indeed, most codes must be well understood because they're the basis of human communication" (Charles, 2000, p. 5) and in essence, information encoding is a process that complements decoding. In this sense, as a process corresponding to decoding, the purpose of information encoding aims to enable specific receivers to rebuild specific receptive relation, while information decoding is the rebuilding of receptive relation. If the receiver is a human agent, the rebuilding of receptive relation is the process of understanding or its product for further understanding. If the receiver is a machine agent, the rebuilding of receptive relation points to the execution of a program.

By understanding information as receptive relation, we can gain a clear understanding of the relationship between information encoding and information: strictly speaking, information encoding is not only about code conversion, but also not about the structural or orderly representation of information itself. Instead, it involves the materialization and idealization of information as receptive relation that is the condensation of the effects of receptive interactions. Memory, records, along with data generated in various other ways are all information encoding as materialization and idealization of information as receptive relation. Therefore, there are two fundamental types of information encoding: material encoding and ideational encoding of information. The former typically being like genes in DNA, biological memory along with electric signals etc., while the latter typically like concepts along with symbols as abstract summarizing products from natural kinds Consequently, with the development of information technology, information encoding has apparently evolved from material encoding to ideational encoding, and further to digital encoding as a special ideational information encoding. Material encoding of information has reality in operations but is constrained by the experiential space constituted by materials; in contrast, ideational encoding of information offers a broader logical space but is relatively detached from matter and energy. This highlights the unique position of information encoding that bridges the gap between material and ideational encoding. With the development of ideational encoding of information, a particular form emerged in philosophical exploration, which is digital encoding of information (bit).

In the development of information encoding, the earliest form was material encoding of information. It was only when receivers evolved consciousness based on material encoding of information and subsequently developed abstract generalization abilities on the foundation of consciousness that ideational encoding of information emerged. Since the advent of ideational encoding of information, a foundation was laid for the development of ideational systems based on ideational information encoding. Through the bidirectional cycle formed between material encoding and ideational encoding of information, humans gained a holistic guan zhao of ideational systems on the establishment of receptive relation and increasingly complex relationships between higher-level ideational systems and receptive relation as information.

5.2 Material Encoding of Information

In the real world, there is a widespread presence of both the informatization of matter/energy and the materialization of information, operating in a bidirectional cyclic mechanism. Among these, the material representation encoding of information is at the core of the materialization of information. In the context of the receptive relation understanding of information, the material encoding of information is the materialization of receptive relation. This means that encoding information in material form involves translating receptivity effects into material codes, effectively leaving a general memory trace (engram) of receptive interactions. Similar to human memory, the encoding of information in material form aims to allow specific receivers to rebuild corresponding receptive relation. Therefore, relative to naturally existing primary sources of information, the material encoding of information is a secondary source. These contents about information encoding can only be recognized in the receptive relation understanding of information.

5.2.1 Biological and Physical Encoding of Information

Due to the fundamental differences between information and matter/energy, the development of information encoding has followed a process opposite to natural evolution. Natural evolution progresses from inorganic to organic, while the development of information encoding has moved from biological encoding to physical encoding.

As a process of receptive relation, the traces of the effects of organismic receptive interactions are the earliest natural information encoding. When natural information decoding develops based on the foundation of natural information encoding, a bidirectional cyclic mechanism of natural information encoding and decoding is developed, and a typical example of this is the biological process of DNA inheritance. In processes involving only organismic receptivity, the source and receiver of information have not yet differentiated, which is mainly why material encoding and decoding do not involve typical signals. The material encoding of information initially occurs naturally, with genes in biological organisms being the classical example of information encoding. Therefore, the genes in DNA can be seen as a nearly perfect product of the integration of matter/energy as organism and information in the natural world. In the material encoding of information, as the earliest complete form, biological genes are both very typical and uniquely advantageous.

Due to the receptive relation nature of information, in order to process and utilize information and carry out information production, there must be material encoding of receptive relation. In the evolutionary process of information, the materialization of receptive relation, as a typical form of information, between receivers and sources of information forms the basis for the development of memory. After the differentiation of receivers and sources of information, there are two fundamental ways in which matter/energy and information interconvert. One is the bidirectional conversion between the informatization of matter and the materialization of information, with the DNA of plants being a typical example. The other is the encoding and decoding of information, with biological and physical memories respectively in the human brain and the central processing unit (CPU) being typical forms of information encoding. Biological genes in DNA are not only a result of materialization but also a typical form of informationization. Memory is no longer a simple trace left by the effects of receptive relation process; it can be both material encoding coded by specific receivers thereby stored in material form, and with it, certain receivers can rebuild specific receptive relation as information.

The purpose of information encoding is to enable specific receivers to rebuild specific receptive relation, and for this reason, receivers must have the capability to decode the information. "Encoding only counts as successful if decoding is in principle possible" (Timpson, 2013, p. 5). If a specific receiver cannot decode, it means that the information encoding has failed. Whenever there is a receptive interaction, there will be receptive interaction effects, which may leave traces, but these traces are not memory or information storage. Purely physical interaction effects always leave traces, but they are unrelated to information and information encoding. The fundamental difference between information encoding and traces lies in the fact that traces are just remnants of effects, while information encoding is a mechanistic information storage, based on which corresponding receptive relation can be rebuilt by specific receivers. Memory is the storage of sources based on that a specific receiver can rebuild specific receptive relation in a way.

As information encoding, memory is the materialization form of information that represents receptive relation. It falls under the materialization of information as receptive interaction effects. Memory is an important area of study in the relationship between information itself and material encoding of information. Through the study of memory, we can clarify the relationship between information and its physical encoding.

Physical imprints are not memory; memory is always informational. Memory is information encoding, and this is the standard for distinguishing traces of physical interaction effects from memory as information encoding. In terms of information, physical imprints are only potential physical attributes that could become information sources, but they can only be potential original sources before certain receptive relation is generated between a specific receiver and them as information sources. Memory, on the other hand, is not only the physical encoding of information but also the potential secondary source that specific receivers can rebuild corresponding receptive relation with.

Since information originates from the organismic receptivity, information encoding initially can only be biological encoding. It's only when the development of information receivers reaches a level of conscious information encoding capability that physical encoding of information becomes possible. One of the most primitive forms of physical encoding of information is finger counting, where fingers are encoded as numbers. When the quantity being counted exceeds the range of finger counting, more advanced methods like knot tying and rope counting develop. In the age of computers, we have seen the development of physical encoding of information, from punched paper tapes to modern USB drives.

Biological encoding of information creates the conditions for the differentiation of information receivers and sources, providing essential foundations for the development of sensory receptivity and information evolution based on organismic receptivity. However, it is limited to a relatively narrow biological domain. Inorganic physical encoding of information opens up vast development possibilities for the materialization of information and the informationization of matter and energy, expanding the realm of information encoding from the biological to the inorganic world and enabling the transition between biological and inorganic physical encoding. One significant aspect of this development is the transition of information encoding from object encoding to energy encoding.

5.2.2 Object Encoding and Energy Encoding of Information

The material encoding of information has undergone a transition from object encoding to energy encoding, highlighting the need for specific distinctions between matter and energy in the field of information. This reflects the unique connotations of Wiener's statement that information is neither matter nor energy.

The material encoding of information is relatively concrete and intuitive. These can be further divided into two basic types: objects and energy encoding of information. Both of these are forms of material encoding, but due to the fundamental differences between objects and energy in terms of movement, they play significantly different roles in the process of information. For instance, in signal formation and transmission, there is a vast difference in speed and efficiency between encoding information as energy and encoding it as objects. The transmission speed of drum signals is described as being "a messaging system that outpaced the best couriers, the fastest horses on good roads with way stations and relays" (Gleick, 2011, p. 13). Because encoding signals through drumbeats involves a conversion from objects to energy as the carrier. Drumbeats as signals propagate at the molecular level of energy, while optical signals propagate at the subatomic level of energy, representing a complete form of energy.

With the development of information technology, signals have gained increasing importance in information processing, and the nature of signals depends on their composition. The development from electronic computers to quantum computers, and their elevated status, is closely related to the fundamental differences in signal composition.

The development of encoding information as objects and energy establishes the material foundation for the ideational encoding of information. The more advanced the development of encoding information as objects and energy, the more closely it is associated with the ideational encoding of information.

5.3 Ideational Encoding of Information

In the current research on information encoding, there is an intriguing phenomenon: on one hand, most of the research focuses on the physical encoding of information, with little consideration for the ideational encoding of information, which is the inevitable result of limited understanding of information. The importance of research on ideational encoding of information can be seen from the ideational occlusion in understanding information. Based on the receptive relation understanding of information, ideational encoding of information appears to be an entirely new field of study. On the other hand, ideational encoding of information is basically blank, but digital encoding of information involved in digitization is precisely a special kind of ideational encoding. However, it has not been studied as ideational encoding of information, but has instead become the source of the most serious misunderstanding of information.

Since "bit" is not understood as ideational encoding and it is not matter or energy, its role and relationship in information encoding are confusing, and it is also widely misunderstood as information itself. In the systematic research on information encoding, clarifying the nature and relationship of digital encoding of information relative to analog encoding of information is of great significance.

In the receptive relation understanding of information, ideational encoding of information is the idealization of receptive relation. The idealization of receptive relation is that information as receptive relation, or the effect of receptive interaction, is coded into ideas including concepts, symbols, pictures, and even experiences etc. by receivers, that is the ideational condensation of receptive interaction effects. A typical example is generating concepts from concrete natural kinds through abstract generalization. The transition from materialization to idealization of receptive relation signifies a fundamentally significant historical change in the development of information encoding, thereby constituting a fundamental distinction between two types of information encoding.

Ideational encoding of information differs from physical encoding of information in a crucial way: it must be converted into corresponding material encoding to be grounded, because only based on their material encoding can specific receivers rebuild the corresponding receptive relation. From this, we can see that on the one hand, code conversion in information-related activities is undoubtedly encoding, but it is not information encoding; on the other hand, ideational encoding itself cannot be a source, and only further conversion into material encoding can be the secondary sources.

As the highest level of development in information encoding to date, ideational encoding of information is built upon two fundamental foundations. One is the corresponding material carriers. All ideational encoding of information has corresponding material carriers, and without exception, ideational encoding finds its realization in corresponding material entities. The second is the corresponding social foundation. On one hand, ideational encoding of information is built upon the foundation of physical encoding of information, so there will be a process from nature to society. On the other hand, concepts are products of social development, and during this development process ideational encoding becomes more and more obviously formed or developed by convention, that is, in a socially agreed manner. From this, we can see an important fundamental point: human concepts form a vast and complex system, among which only the most basic concrete concepts are information encoding, while the majority of concepts are higher-level concepts built upon the foundation of ideational encoding of information. This applies to information encoding at all levels and is crucial for understanding ideational encoding of information and, by extension, the receptive relation understanding of information.

Ideational encoding of information, as a process of idealization, has undergone a development from experiential encoding of information to graphic (including from concrete representation to abstract graphics) encoding, and further to conceptual and symbolic encoding. This progression involves varying levels of complexity in information encoding and processing mechanisms.

As a new research field, ideational encoding of information holds special significance for systematic research on information encoding. However, due to the complexity of dealing with ideational aspects in information encoding research, it becomes especially intricate in its early stages. Due to space constraints, this discussion primarily focuses on the ideational encoding of information as it pertains to conceptual encoding and symbolic encoding, briefly summarizing earlier stages of ideational encoding as preliminary conceptual encoding of information.

5.3.1 Pre-Conceptual Encoding of Information

As receptive relation, ideational encoding of information, which emerges at the higher level of development of information encoding, differs both from memory and sensation. It represents a higher-level product of information development based on the receiver's perceptual capabilities. In Dretske's discussion of information, a specific example helps illustrate the sensory genesis of information experiential encoding:

> Is all the information embodied in the sensory representation (experience) given a cognitive form? No. You saw 28 people in a single brief glance (the room was well lit, all were in easy view, and none was occluded by other objects or people). Do you believe you saw 28 people? No. You didn't count and you saw them so briefly that you can only guess. That there were 28 people in the room is a piece of information that was contained in the sensory representation without receiving the kind of cognitive transformation (what I call digitalization) associated with conceptualization (belief).
>
> (Dretske, 1983)

The receptivity of receivers implies certain receptive capabilities, and the information they obtain depends on the level of their receptive abilities. In the case of seeing 28 people in a room, with just a quick glance, most people cannot accurately perceive the exact number, but for artificial intelligence, it's a simple task. For humans, if there are only five people in the room, the receptive outcome has a meaningful difference. A typical adult should be able to accurately perceive the exact number with just a quick glance; while a child who doesn't possess such visual receptive abilities would still have to guess. This illustrates the development of ideational encoding of information and the different levels it takes on as it evolves.

Built upon the foundation of sensation, the initial forms of ideational encoding were primarily experiential encoding and graphic encoding, constituting the two fundamental forms of pre-conceptual encoding of information.

Based on sensory experiences, the earliest forms of ideational encoding involved the establishment of associations between sensations, forming the initial experiential encoding of information based directly on memories. Information experiential encoding is directly built upon the foundation of receptive relations, composed of various receptive relations established in the environment and their interconnections. However, not all experiences are information encoding; information experiential encoding includes only the basic experiences. As the level of associations between sensations develops, experiences become more complex, and these complex experiences form the intuitive experiential system based on information experiential encoding. This represents the first important node in the development of information encoding, forming an experiential whole based on the foundation of basic experiential encoding. Experiences have various elements, and their representations constitute the graphic encoding of information.

Graphic encoding of information initially establishes on the basis for sensation about object shapes, with memory of these shapes evolving into a stage of representation. graphic encoding of information holds two significant implications. Firstly, it leads to the development of imagination based on graphic encoding, where imagination is no longer simply the product of graphic encoding but is a system of graphics generated on the basis of information graphic encoding. This represents the second important node in the development of information encoding, vividly demonstrating the relationship between ideational encoding of information and the ideational systems built upon it. Imagination and its products are not graphic encoding but rather a graphical relationship system established on the foundation of information graphic encoding, often exemplified in the study of geometry. Secondly, the abstraction of specific graphic encoding leads to the development of conceptual encoding of information. It is conceptual encoding of information that marks a significant turning point in the development of information systems, laying a crucial foundation for human civilization.

5.3.2 *Conceptual Encoding of Information*

The conceptual encoding of information is a typical form of ideational encoding of information, which is the product of the development of ideational encoding of information to a high level of generalization.

In receptive relation as information, the most fundamental is the receptive relation with a specific material-energy existence as the source. As an abstract generalization or conceptualization of this receptive relation, conceptual encoding of information represents a revolutionary breakthrough in information encoding. Material encoding of information is concrete; conceptual encoding of information is abstract. To advance our understanding and thinking based on information to higher levels, we cannot do so without abstract conceptual encoding. Without conceptual encoding, actually up to all ideational encoding, the development of information cannot transcend the limitations of direct material attributes and thus break free from the limitations of concrete things. It is the conceptual encoding of information that allows us to surpass the receiver to transcend the material limitations of material encoding of information, not only to establish a broader logical space based on the foundation of experiential space, but also to lay the foundation for the development of higher-level systemic theories, as well as meaning systems and value systems within information encoding.

The conceptual encoding of information is a mature form of ideational encoding of information, and thus, it encompasses the most extensive content related to conceptual research and its connection to information research. Dretske noted,

"Although we will continue to insist on the distinction between the concept of meaning and the concept of information, we will later argue that the idea of information is more fundamental. Meaning is generated by the way information is coded.

(Dretske, 1982, p. 248)

Information encoding, especially abstract encoding of information is related to meaning. However, under the conditions of sentient beings with a knowledge background, the receptive relation itself already has a certain level of assignment based on property endue of the receiver. This involves the important role of conceptual systems in the process of establishing receptive relation and the relationship between conceptual systems and information conceptual encoding. The study of concepts not only provides rich resources for thinking about information conceptual encoding but also allows for a deeper understanding of the nature of concepts.

The concept, which can be generated from logical deduction, is an infinitely expandable set. The encoding of the concept of information is just a very small basic part that is directly related to receptive relation as information. This represents the third important node in the development of information encoding: establishing conceptual systems based on conceptual encoding, the most typical form of ideational encoding of information. The qualitative understanding of objects established by this milestone is crucial. It can better explain the relationship between information conceptual encoding and conceptual systems, especially the relationship among concepts of different natures.

Compared to the set of concepts, not all concepts are the conceptual encoding of information; only those that serve as generalizations of receptive relation are the conceptual encoding of information. Regarding sensory receptivity, when a person sees tree leaves and perceives them as "green" in the visual system, the "green" is the effect of perceiving specific wavelengths of light reflected by tree leaves, and "green" as receptive interaction effect is information itself. However, higher-level concepts derived from these information codes through logical abstraction, "green," "red," and "blue" as the products of specific color receptive effects generalized by the receiver, are conceptual encodings of information, while higher-level concepts derived from these information conceptual encodings based on logical relationships, such as "color" and "spectrum," are no longer information encoding; they are further abstraction of "red,", "green," and "blue" and so on as conceptual encodings based on logical relations. Therefore, anything derived from further abstraction of information conceptual encoding is no longer the conceptual encoding of information but rather higher-level concepts that are deduced from concrete conceptual encodings, they themselves are not the encoding of information as concrete receptive relation. The higher the level of concepts derived from information conceptual encoding, the further they are from the corresponding receptive relation. Consequently, the importance of information conceptual encoding becomes evident, as does its intrinsic connection with human knowledge and the nature of theories.

5.3.3 Symbolic Encoding of Information

Just as the abstract development of information representation encoding leads to concepts, further abstract development of information conceptual encoding leads to symbolic encoding of information. This is a process that constantly divorces from concrete conditions. Symbolic encoding takes the abstraction of ideational encoding to the extreme. By removing all concrete experiential conditions, symbolic encoding not only allows for formal manipulation but also facilitates machine simulation. This marks the fourth crucial node in the development of information encoding: establishing a symbolic system based on the most abstract form of ideational encoding of information. This node has fundamental importance, as it is not only relevant to formal sciences like mathematics and logic but is also evident in the development of information technology characterized by digitization. It is this development that not only expands the richness of information hierarchies but also directly influences how people understand information. When information encoding is treated as information itself, there can be various understandings of information, particularly when symbolic encoding is involved. Hence, the significance of these understandings mostly lies not in the information itself but in the encoding of information.

As information encoding progressed from physical encoding to ideational encoding, it developed various levels. Symbolic encoding as a specialized encoding form, not only belongs to one of the two different fundamental levels of information encoding, but also includes different ways of coding itself, one of which is "numerical coding," or "digital." The proponent of numerical encoding, Tribes, suggested, probabilities are treated as a numerical encoding of a state of knowledge. One's knowledge about a particular question can be represented by the assignment of a certain probability (denoted p) to the various conceivable answers to the question. Complete knowledge about a question is the ability to assign a zero probability ($p = 0$) to all conceivable answers save one. A person who (correctly) assigns unit probability ($p = 1$) to a particular answer obviously has nothing left to learn about that question. By observing that knowledge can be thus encoded in a probability distribution (a set of probabilities assigned to the set of possibilities), we can define information as anything that causes an adjustment in a probability assignment.

(Tribes & McIrvine, 1971)

In essence, this definition of information is not substantially different from Shannon's. However, from the perspective of understanding information encoding, it offers a different view. Therefore, Yixin Zhong pointed out, "Although this definition may seem quite different from Shannon's definition, they are essentially the same" (Yixin, 2013, p. 61). This interpretation is essentially related to Shannon's definition of information, but they are only consistent with the existential aspect of Shannon's information understanding rather than the signal aspect. In this sense, they both have important implications for understanding information encoding.

The "numerical encoding" discussed by Tribes may seem connected to the source-preexisting understanding of information, but since it involves symbolic encoding of information, it is associated with the receiver and actually has the nature of the receiver-assignment understanding of information. Therefore, its meaning in understanding information is not too unique, but it has great value in its contribution to the study of information encoding: it opens the door to and advances research on the unique field of digital information encoding.

In terms of fundamental nature, numerical encoding of information belongs to ideational encoding of information, but due to the unique importance of digital encoding, not only is it necessary to study digital encoding as a type of information encoding alone, but also because it has theoretical and practical integration significance, specialized research on digital encoding has more important aspects than physical encoding and ideational encoding: it involves important distinctions between analog encoding and digital encoding of information.

From the perspective of the receptive relation understanding of information, digital encoding is a special form of information encoding that represents the developmental transition from material encoding to ideational encoding. Due to its significant advantages over analog encoding, digital encoding has experienced explosive growth. Some literature even describes this development as the "world's digital transition" (Béranger, 2018, p.141). It signifies neither physical encoding of information nor the general ideational encoding of information, but rather a unique form of ideational encoding that can be directly associated with physical encoding of information. It is precisely because physical encoding and ideas are inherently associated that a complex relationship between ideational encoding of information and human ideational systems has been unfolded.

5.4 Information and Ideational Systems

The understanding of information as receptive relation means that the colorful world displayed in sensory receptivity is actually an information landscape as an effect of receptive interaction, and what level of holistic landscape can be established for the holistic existence of the physical world, or composed of source, cannot rely solely on the senses, especially in complex situations, it must rely on the ideational system established based on information encoding. This implies that although the conceptual system based on information ideational encoding is not information itself, with the deepening of receptive interaction, it plays an increasingly important role in the establishing of receptive relation increasingly deep.

5.4.1 Information is Neither Matter and Energy nor Any Form of Ideas

Exploring the ideational encoding of information on the basis of its physical encoding has significant implications. It is not only closely related to the further in-depth study of knowledge as a conceptual system but also to the systematic deepening of information understanding.

Regarding the relationship between information and concepts, it can be seen very clearly from the receptive relation understanding of information. In fact, not only neither matter nor energy, but also ideas are not information itself. While ideas are neither matter nor energy and do not have an inherent connection with matter and energy, they cannot be information itself either, and otherwise, it would imply that information can exist independently of matter and energy. Conversely, this underscores that ideas are not information itself but also that information is not relation between ideas. Information is not a static existence but a dynamic process, existing solely as receptive relation in dynamic processes. As receptive relation, information is not only empirical but also at the most fundamental level of experience.

From the fact that information is not any sense of information encoding, it becomes apparent that information is also not the ideational encoding of information and further raises a new issue: neither matter nor energy doesn't constitute a comprehensive negative provision of information because there are also ideas that are neither matter nor energy. It is precisely from the ideational encoding for information that leads to a natural conclusion: information is also not conceptual encoding. Information is neither matter nor energy nor any form of concept. To regard pictures and data as information itself represents an information encoding understanding of information. The most typical example of this is understanding information as bits in digital encoding. The understanding of information in bits, as exemplified by John Wheeler's "it from bit" (Wheeler,1990) distinctly illustrates this. As a matter of fact, even though information is bits, not all information is digital (Pintar and Hopping, 2023). Going even further, interpreting knowledge as information equates to regarding a conceptual system built upon information encoding as information itself.

Given that the ideational encoding of information is also not matter or energy, there is a need to impose a more specific constraint in line with Wiener's negative definition of information understanding. Information is neither matter nor energy, nor is it ideas, meaning it is not information encoding in any sense or ideational systems built upon it. Information is not any physical encoding like genetic or electrical signal, nor is it the analog or digital encoding of data and knowledge, and it is not the ideational systems or any ideas based on information encoding.

Clarifying the relationship between information and its ideational encoding actually implies the establishment of a series of important new relationships. The clarification of the relationship between information and data provides a deeper foundation for data research. The separation and clarification of knowledge from information attachment provide a deeper foundation for epistemological studies, marking the beginning of the theory of knowledge based on information. The most direct area of research here is the relationship between information and the ideational encoding of information, as well as the relationship between ideational systems based on information encoding.

5.4.2 Ideational Encoding and Ideational Systems of Information

One important reason why research on ideational encoding for information has not received due attention is the complexity of the relationship between idealization of information and ideas. If we directly link ideas with information, it will result in various ideas and ideational systems, being regarded as information and the higher-level the ideational system, the more difficult it is for people to accept it as information itself. In reality, this leads to ideational veils in information understanding, and the comprehension of the relationship between information and ideas becomes even more problematic as we delve into more complex ideational systems. Consequently, research on information and ideas has always been conducted separately while being close at hand. The new foundation provided by the receptive relation understanding of information is essential for comprehending the relationship between ideational encoding of information and ideational systems, providing a foundation for clarifying this relationship. Starting from a receptive relation understanding of information, we can identify significant intellectual resources closely related to information research.

Based on the ideational encoding of information, we can derive not only higher-level ideas but also establish various ideational systems. Even though they have departed from information encoding, they are indispensable for building deeper receptive relation. By relying on building increasingly higher-level ideational systems established based on information encoding, the objects of receptive relation expand from a single leaf to a tree, a forest, a mountain, or a planet etc. and thereby constitute a bidirectional cyclic information mechanism.

Concerning the relationship between ideas as information encoding and ideational systems, we can gain a vivid understanding from a specific example involving ideas and ideational systems in Dretske's work based on the receptive relation understanding of information. In the development of receivers, such as humans, the relationship between information as receptive relation and ideational systems based on information ideational encoding is exceedingly intricate. Humans interact with the world solely through their sensors, but they often become lost in their sensors due to rational constructs built upon sensory foundations. In terms of receptive relation, when a person sees a leaf, it appears as the "green" effect in their visual system, as a result of perceiving specific wavelengths of light reflected by the leaf – an effect (relation) as part of the receptive interaction. This "green" effect, when retained as a memory in the brain, becomes a specific encoding of information in the form of biological memory. Forming the concept of "green" through abstract thinking is the ideational encoding of information. Using "G" to represent green is symbolic encoding within the information ideational encoding. Biological memory serves as a specific encoding of information, while both "green" as an ideational encoding and "G" as a symbolic encoding are abstract encodings of information. Higher-level abstract concepts derived from information ideational encoding no longer belong to information encoding but encompass all concepts within the established ideational systems.

In discussions about analog encoding and digital encoding of information, Dretske provides a helpful example that deepens our understanding of the relationship between information conceptual encoding and ideational systems. Dretske explains,

> Perception is a process in which incoming information is coded in analog form in preparation for further selective processing by cognitive (conceptual) centers. The difference between seeing a duck and recognizing it as a duck (seeing that it is a duck) is to be found in the different way information about the duck is coded (analog vs. digital).
>
> (Dretske, 1983)

Rather than merely illustrating the relationship between analog encoding and digital encoding, this example vividly illustrates the relationship between information ideational encoding and the ideational systems built upon it. When an actual duck is observed, the effect of seeing it is information as receptive relation. The formation of the concept "duck" or any other form is the conceptual encoding of information, while "recognizing it as a duck" implies distinguishing it from other objects and, strictly speaking, is closely related to the conceptual systems built upon concepts, including "duck." The more complex the object being recognized, the more evident this becomes. Simpler objects may be recognized based on concepts, while more complex objects require corresponding conceptual systems. Therefore, research on conceptual systems based on information encoding not only involves the relationship between information and knowledge, theories, and so forth, but is also directly related to the deepening of research on information as receptive relation.

In the context of understanding the relationship between the ideational encoding of information and the ideational systems built upon it, we can not only gain a clearer insight into the issues surrounding information comprehension but also unveil its deeper significance. Throughout the development of information research, various approaches to understanding have made indispensable contributions, even when some important viewpoints are challenged and counter-challenged, they collectively drive the advancement of information research.

As receptive relation, the association between information and concepts primarily exists in two significant aspects: firstly, information ideational encoding as the generalized reflection of the existence of concrete individual entities; secondly, the role of ideational systems based on information ideational encoding becomes increasingly important in the deepening of the establishment of further receptive relation. Another example discussed by Dretske illustrates this point clearly. After all, any incoming signal that carries the information that *s* is water carries (nested in it) the information that *s* has oxygen atoms in it (since there is a lawful regularity between something's being water and its having oxygen atoms in it). The answer to this question is, of course, that the child has not developed a sensitivity to the information that *s* has oxygen atoms in it just because the pupil has been taught to respond positively to signals all of which carry that information. (Dretske, 1983) What is clearly involved here is the crucial role of the ideational system of the receiver in the establishment of receptive relation, with the key mechanism being what is often referred to as the theory-ladenness of observation.

Because ideational systems based on information ideational encoding, knowledge and theories, for instance, are neither information encoding nor information itself, we can see both sides of an important coin. On one hand, ideational systems established based on ideational encoding of information give humans the freedom to transcend specific receptive relation, allowing them to roam in the realm of logical space. Thus, human cognitive space can be limitless, as it can establish, in principle, infinite relational systems based on information encoding. This enables human thinking to expand into any imaginable domain and, through formal deduction, extend into logical spaces beyond human imagination. On the other hand, within ideational systems, the higher the level of concepts, the further they are from information as receptive relation. Information ideational encoding is concrete, and in higher-level concepts and their systems derived from it, foundational concrete concepts are closer to receptive relation, while concepts deduced from multiple conceptual levels are further from information, eventually becoming entirely unrelated to the corresponding information as receptive relation. This explains why there are abstract concepts and theories detached from reality.

The detachment between theory and practice is not only evident in human activities but also exists at a deeper level of information, extending to the encoding, including memory as biological encoding of information. In fact, even in the case of memory based on corresponding proteins, information ideational encoding, especially the ideational systems built upon it, can introduce unforeseen uncertainties, often closely related to the thought stipulations underlying the ideational system as the foundation. As more complex memories involve ideational systems based on information ideational encoding, the rebuilding of corresponding receptive relation becomes a process that must be realized through understanding and interpretation, leading to the phenomena of both forgetting and unintentional manipulation of human memory.

5.5 Conclusions

Due to the information encoding understanding of information, the understanding of information encoding has always been intertwined with the concept of information itself. The receptive relation understanding of information makes systematic research on information encoding possible. Due to the nature of receptive relation understanding of information, information encoding forms the foundation for utilizing information, as well as information processing and production.

As the result of the condensation of receptive interaction effects, information encoding is the materialization and idealization of information. As materialization of information, physical encoding of information typically involves material aspects such as genes in DNA and biological memory. Genes, as an example of physical information encoding, have the capacity to rebuild the organismic receptive relation. They are not only the result of informationization of matter and energy but also a typical form of information materialization. The characteristic of memory is that a certain receiver can directly rebuild specific receptive relation with it. The physical encoding of information has gone through a development process from object encoding to energy encoding, which is an important reason why the field of information research needs to make a specific distinction between matter and energy. Both the object encoding of information and the energy encoding of information are material encodings, but they are completely different in terms of signal formation and transmission speed and efficiency. The more the physical encoding of information develops to a higher level, the more closely it is related to the ideational encoding of information.

Ideational encoding of information is a brand new research field. The only thing that involves ideational encoding of information in existing research is digital encoding of information, and it is not studied as ideational encoding. As a new research area, ideational encoding of information provides a new theoretical foundation for a deeper understanding of information, data, knowledge, and their interrelationships. Research into ideational encoding of information not only aids in a better understanding of information encoding but also contributes to a greater understanding of the formation and development of human concepts.

Ideational encoding of information is grounded upon corresponding material carriers and social foundations and has evolved from experiential encoding of information to graphic encoding, and further to conceptual encoding and symbolic encoding. Ideational encoding represents a revolutionary breakthrough in information encoding, among which, the conceptual encoding is a product of the generalization of natural kinds that exist concretely, so only those concepts that are products of the abstraction of receptive relation are conceptual encodings of information. The more developed the conceptual system, the more concepts there are that are not information conceptual encodings. As a product of the development of information conceptual encoding abstracted from specific empirical conditions, symbolic encoding of information can form formal operational systems for machine simulation, paving the way for the development of electronic computers and intelligent machines. Digital encoding of information is a specific form of development from material encoding to ideational encoding. It can be symbolic encoding of ideational encoding of information and can also directly convert into material encoding of information. Digital encoding of information has a special position in the relationship between material encoding and conceptual encoding. Digital encoding of information not only reveals the ideational encoding of information but also links material encoding of information and the physical world closely together, forming not only an integrated information encoding system but also a strong connection between information encoding and the physical world.

The material encoding of information is concrete, whereas ideational encoding of information gives rise to abstract conceptual encoding and symbolic encoding. The abstract nature of conceptual encoding allows it to transcend direct material limitations and break free from the confines of specific objects. Ideational encoding of information allows receivers to establish a broader logical space on the foundation of experiential space, laying the information encoding groundwork for the development of increasingly hierarchical conceptual systems, thereby establishing systems of meaning and values. Concepts, especially symbols as ideational encodings of information, can be grounded in memory etc., as physical encodings on the one hand, and exist as abstract concepts on the other. Based on this abstract concept, ideational systems can be established, thereby possessing the existence of a logical form that is relatively independent of material, For instance, further concepts or conceptual systems can be deduced through logical reasoning based on externalized text and symbol relationships, even when they are not stored in memory. In fact, ideational encoding of information represents a significant information foundation that elevates information processing from relatively simple information processing to relatively complex cognitive production.

As the materialization and idealization of receptive relation, information encoding, especially the ideational encoding of information, involves important relationships between information, concepts, and conceptual systems. Because it not only provides a realistic basis for the establishment of a conceptual system through conceptual encoding but also provides a specific mechanism for the direct docking of a conceptual system and the physical world through digital encoding. As a result, the role of ideational encoding of information has become increasingly important in the field of information research. With the deepening development of receptive relation as information, the ideational encoding of information and the ideational system based on it are becoming increasingly important.

Research on the ideational encoding of information shows that information is neither matter nor energy nor any form of idea. This is of great significance for deepening the understanding of the relationships between information, data, and knowledge. On the one hand, in the context of the relationship between information and data, clarifying this relationship provides a more profound informational basis for data research. On the other hand, in the context of the relationship between knowledge and information, dissecting and clarifying this relationship provides a deeper informational foundation for epistemological studies, marking the beginning of knowledge research based on information.

Through the lens of information ideational encoding, it becomes evident that ideational systems based on ideational encoding of information are not mere reflective mirrors of objects; instead, they provide a holistic guan zhao for the establishment of receptive relation. Here we can get a better understanding of the phenomenological perspective. "When we perceive an object, we always experience more than what is intuitively presented." (Dan, 2019, p.11) The simpler the receiver's ideational system, the more objectively receptive relation can reflect the characteristics of the source, but without deep understanding, it is impossible to have a higher level and more rational holistic guan zhao. The more complex the receiver's ideational system, the higher the content level of the established receptive relation, and the higher the understanding and guan zhao of specific things, but the greater the possibility of over-understanding.

In the deepening development process of receptive relation as information, the ideational system based on information ideational encoding contributes to the holistic guan zhao in the process of establishing receptive relation, making receptive relation as information develop from simple receiver property endue to increasingly complex receiver semantics in the establishment process. The receiver property endued in the process of establishing receptive relation refers to the receiver intrinsically enduing its properties to receptive relation, while the receiver assigning intrinsically refers to a receiver with understanding capabilities intrinsically endue their understanding to receptive relation. Digital data is digital encoding of information. It can be information encoding affected only by the receiver's characteristics, or it can be information encoding with high-level receiver enduing intrinsically. Receivers enduing intrinsically are increasingly present in the establishment of receptive relation and play a progressively deeper role in the receptive relation established between receivers and secondary sources.

In summary, for a clearer understanding of the relationship between information and matter/energy, research into information encoding is an essential foundation. However, there are deeper internal connections between information encoding and its complex relationship with information and matter/energy. It is necessary to further delve into the level of information principles related to internal connections with matter and energy in order to have a deeper grasp.

6 The Principle of Identity for Information-Matter in Operation

6.1 Introduction

To a large extent, the process of human understanding of information is also a process of clearing the fog of the relationship between information and matter. The fog surrounding the relationship between information and matter is particularly evident in the "information conservation" dilemma.

In material science, scientists not only naturally assume "information is physical," but in the realm of natural sciences, it is clear that information processing follows the conservation laws of matter/energy. Therefore, the nature itself of information implies information conservation. Matter and energy are conserved; matter can only change its form, and energy can only be transformed. Matter and energy conservation in the material world is self-evident. Matter and energy can neither be created nor destroyed. However, closely related to this, the conservation of information in the sense that it cannot be created or destroyed poses a fundamental challenge to the understanding of information.

Among the viewpoints that lead to fundamental challenges and even paradoxes in information understanding, "information conservation" is not only the most obvious and acute but also the richest in meaning and most explanatory. In the study of information in physics, the idea that "information is physical" is natural and thus taken for granted, as otherwise, the subject of study would not be physics. This understanding is even more typical in the field of information biology. However, in the logically sound understanding of physics and information biology, there is an evident problem: it inevitably leads to the conclusion that information, like matter and energy, is conserved. "Conservation of information" means that "at the fundamental level information cannot be destroyed, but only discarded," and "its validity is often upheld as one of the most fundamental laws of physics" (D'Ariano et al., 2017, pp. 187, 170). As far as information encoding is concerned, as long as the physical encoding of information is considered as the information itself, there is no problem with "information conservation." However, this contradicts the fundamental fact in everyday life that the creation of information implies making something out of nothing. As the highest-level consumers of information, everyone can directly prove that, even if not entirely, information can be non-conservative. And as long as there is a phenomenon of non-conservation of information, the law of information conservation cannot hold.

DOI: 10.4324/9781003484851-6

Just as "information is physical," the "information conservation" principle also stems from the subtle relationship between information and matter. The logical basis for this lies precisely in misunderstanding information as matter, thus establishing its conservation characteristics through the endowment of matter. This is indeed a very subtle and intriguing logical relationship: its arguments can be derived directly from the conclusion it aims to prove.

The deeper the fog, the deeper the principals involved. Using "information conservation" as a sign, the multiple layers of fog in the relationship between information and matter have deeper principled roots. Delving deep into the relationship between information and matter not only clarifies the fog at its roots but also reveals an important and special fundamental principle of information.

The paradoxical dilemma caused by the contradictory nature of the problem is often the gateway to a higher level of understanding. On the basis of the receptive relation understanding of information, the higher-level entrance implied by the information conservation dilemma is within reach. As receptive relation, information must ultimately be realized in the physical encoding to enter practical operation. Even the ideational coding of information in the brain must be realized on the basis of biochemistry and bioelectricity. It is the information conservation dilemma that actually constitutes an important guide to an information principle. This can lead to an important fundamental principle involving the inherent connection between information and matter: the identity of information-matter in operation. This unique basic principle of the relationship between information and matter not only reflects the fundamental connection between information and matter but also embodies the uniqueness of information compared to matter, making it of special significance.

From the principle, not only can the clear misunderstanding of "information conservation" be seen, but also the rationality underlying this misunderstanding can be discovered, providing a principled basis for clarifying this fog. Its systematic study can not only clarify various important related cognitive fogs, including the "information conservation" dilemma and issues of information grounding related to the symbol grounding and the quantification of information but also clarify the boundary between quantum physics and classical physics at a deeper level. Moreover, it can more precisely define the mechanistic relationship between information and matter.

6.2 Information and the Operation of Its Physical Encoding

Because receptivity can only be a characteristic that emerges at a certain level of matter development, information as receptive relation ultimately must be built on the foundation of matter. In this sense, information is receptive relation of matter. Even though the source of information can be something other than matter (in fact, it corresponds to a non-existence relative to matter), the receiver of information must be based on matter as its carrier; otherwise, information cannot exist, let alone enter into practical operation. It is precisely for this reason that the assertion "information is physical" is based on the fact that information can only exist and be put into practical operation when it is embodied in matter.

From the understanding that information is receptive relation, on the one hand, the proposition that "information is physical" has its validity; it reflects the operational nature of information, so it should be stated more accurately as: "information operation is physical." On the other hand, in this proposition, the notion that information itself is physical is not only incorrect but even misleading, as it leads to the understanding of information as matter or as a physical entity, ultimately equating information with matter. In the matter-based receptive relation understanding of information, it is not only clear that the physical encoding of information is not the information itself, but it can also be seen at a deeper level that it is a secondary source of information, a way in which information exists in the form of matter. This is because all information processing can only take place through the processing of the physical encoding of information, and all information quantification can only operate indirectly in this way. It is from the operational level of physical encoding of information that another aspect of understanding information becomes clear: the perspective that "information is physical" is an important intermediate step in the understanding of information emerging from the obscurity of matter. This involves entirely new principled relationships. Because it is closely related to the operation of physical encoding of information, this important proposition is not only an important aspect of understanding information itself but also the key to a deeper grasp of the relationship between information and matter, involving the fundamental principle of information-matter.

Based on the understanding of information as receptive relation, the understanding of the relationship between information and matter can delve deeper into the relationship between physical encoding and ideational coding of information. The relationship between these two fundamental forms of information encoding signifies the connection between the worlds of information and matter at a higher level of development. In the receptive interaction between the receivers and sources through signals, one can typically see this relationship unfolding. To establish receptive relation between the receivers and sources through signal transmission, information must be encoded in terms of matter because only matter can serve as a means of transmitting signals. Research on mechanisms such as signal transmission and processing falls within fields like communication science, computer science, and technology. However, research on the properties of the medium serving as the carrier for signals and related fundamental principles belongs to the field of physics. It is for this reason that the complexity of quantum information is not in the information itself but in the process properties associated with quantum observations and processing, that is, the relationship between matter and information generated by the relevant medium in quantum observations and processing.

In the double-slit experiment, quantum phenomena manifest as wave-like properties; in the Wilson cloud chamber experiment, quantum phenomena appear as particle-like properties. If we consider quantum phenomena as the external expression of quantum entities, we would believe that quantum information is a form of matter, or at least see matter as the direct carrier of quantum information itself. This clearly leads to a physical understanding of information, but this contains important rationality about relationships between information and matter/energy.

Because without matter, information cannot be generated or exist, let alone be transmitted, processed, and produced. It is precisely for this reason that information can not only operate in a way akin to matter through operations encoding it into matter physically, but also be measured in an energetic manner based on physical operations. The computation and processing of information must necessarily rely on the encoding of information into matter and energy.

As early as the beginning of the 20th century, Hartley began studying the measurement of information. Regardless of what these physical symbols represent or what meaning they hold, the crucial point is that each physical symbol signifies a choice within a specific set of symbols, which is the encoding of information into matter. Hartley illustrated this with a vivid example: in the sentence "Apples are red,"

> the first word eliminates other kinds of fruit and all other objects in general. The second directs attention to some property or condition of apples, and the third eliminates other possible colors. It does not, however, eliminate possibilities regarding the size of apples, and this further information may be conveyed by subsequent selections. Inasmuch as the precision of the information depends upon what other symbol sequences might have been chosen it would seem reasonable to hope to find in the number of these sequences the desired quantitative measure of information. The number of symbols available at any one selection obviously varies widely with the type of symbols used, with the particular communicators and with the degree of previous understanding existing between them.
>
> (Hartley, 1928)

These choices are realized through operations on physical symbols, leading Hartley to suggest measuring information from a "physical" rather than a "psychological" perspective. It is precisely because the measurement is grounded in matter and energy that information theory provides a quantitative grasp in fields such as communication science.

Both the measurement of information and the quantitative grasp of information theory have significant operational implications for information, at the same time, they have limitations in understanding information. On the one hand, focusing on physical encoding provides a foundation in matter and a physical basis for the operation and measurement of it. On the other hand, because it can be quantified, regardless of what meaning these physical symbols represent, it leaves a fog in information understanding. As a result, it is easy – even under specific conditions inevitably – to lead to an understanding of information in terms of matter and energy. However, in communication science, this understanding of information is not only unproblematic within the discipline, but also offers a more convenient side.

Due to the prevailing material paradigm in the development of natural sciences, research on information initially naturally followed the material paradigm. Since the typical object of study in fields like communication science is signals encoded as information encoding, human scientific research on information naturally started from the physical encoding of information. Communication science certainly involves information, but its object of study is electrical signals, and electrical signals are physical encoding of information. Therefore, understanding information as its physical encoding is not only reasonable but also does not pose a fundamental problem within the field of communication science. It is this non-problematic issue that implies the important principles underlying information and its relationship with matter.

6.3 Principle of Identity for Information-Matter in Operation and Its Physical Boundary

All specific operations related to information, such as encoding, decoding, and measurement, must be carried out through physical encoding. This implies that all information operations are physical processes. It is precisely for this reason that the validity of the statement "information is physical" is not only because the research on information began as the study of physical encoding of information, but also because this proposition has its basis. However, as our understanding of information has developed, it becomes evident that the statement "information is physical" encompasses two tendencies of understanding: one that reduces information to matter and another that leads to the revelation of the physical nature of information operations.

6.3.1 An Operational Principle Inner-Connecting Information and Matter

Since, as receptive relation, the operation of information must ultimately be grounded in physical encoding, therefore, the operation and processing of information inevitably follow the same physical laws, that is, in operation, information and matter/energy exhibit an identity. Precisely because of this, an important fundamental principle can be derived – the principle of identity for information-matter in operation (PIIMO):

In terms of operation, information and matter exhibit an identity.

The principle of identity for information-matter in operation can be represented as:

$$I =O= M$$

Here, "I" stands for information, "M" for matter, and "=O=" indicates operation identity. This principle shows that since as receptive relation, information must be embodied in physical encoding, information operations follow the laws of physical operations. Therefore, at the practical operational level, information and matter/energy follow the same laws, and both have an identity in terms of operation.

The PIIMO suggests that since the operation of physical encoding of information follows physical laws, information operations also follow the principles of physical operations in practice, making the operation of information follows the same principles as those of physical operation and these two operations are equivalent in corresponding fields. This principle has obvious special features, which is why we can see the status of this new symbol "=O=."

In the material paradigm, this identity appears as the coexistence of information and energy – this is precisely why "information is physical" tends to reduce information to matter and energy; while in the paradigm of information, this identity only refers to physical encoding of information – and this is exactly why "information is physical" has rationality in operation. It's precisely because of the special nature of the PIIMO that we can see different natures and characteristics between principles of information and those of matter and energy. Because it deals with a deeper level of the relationship between information and matter/energy, this principle has a different nature than the physical principle, including the principles derived from the physical understanding of information. The unique aspect of the PIIMO is that it pertains to practical operations rather than ontological meanings, i.e., it only has operational significance but does not have ontological significance. This highlights the information-based nature of operationalism and reflects different research paradigms. This characteristic of the PIIMO reflects an important property distinct from the physical principle: the physical limits of the principle of information. Thus, this principle has a physical boundary related to it.

6.3.2   The Physical Limits of the PIIMO

Regardless of whether we approach information from the perspective of the signal-preexisting, the source-carrying, or the receiver-assigning understanding of information, there is an important commonality: it is challenging to fundamentally distinguish between information and matter/energy, as they are either seen as entirely separate or considered indistinguishable. It is the PIIMO that provides a deeper and more fundamental basis for understanding the relationship between them. As receptive relation, information not only exists as real as matter but also arises from its receptivity. However, the process of generating receptive relation through this connection is not inherently distinguishable from matter and energy. It simply requires a deeper understanding of the nature of the PIIMO.

The PIIMO is meaningful in an operational sense, rather than in an existential sense. There is a fundamental difference between operational identity and existential identity: in terms of the effects of principle, the latter is bidirectional, meaning it is mutually symmetric, while the former is unidirectional, leading to symmetry breaking. The PIIMO follows physical laws, but there is no direct regularity connection between physical operations and information in an existential sense; the effects of physical laws are not symmetric. In contrast, the principle of existential identity implies that both information and matter follow the same physical laws bidirectionally, resulting in complete symmetry in the effects of physical laws.

Therefore, the PIIMO has physical limitations: it is confined to the realm of physical encoding and processing of information.

This means, on one hand, from a foundational perspective, the PIIMO can only cover the encoding of information and cannot encompass the receptive relation as information or its ideational coding. Furthermore, it is unrelated to ideational systems based on information encoding. On the other hand, in terms of subsequent processes, the application of the PIIMO cannot extend to the receiver's decoding or understanding during the rebuilding of receptive relation. Its effective scope is limited to the processing of the physical encoding of information; this pertains to signals in communication science, physical encoding in physics (typical in quantum physics) is actually about information physical encoding in the observation process; while in biology it's about processing biological encoding of information. From this, we can see more specific rational factors about operationalism.

Operationalism suggests that we cannot know the meaning of a concept unless there is a way to measure it. "We mean by any concept nothing more than a set of operations; the concept is synonymous with the corresponding set of operations"(Bridgman, 1958, p. 5). With the PIIMO, we can see that there are some rational factors of the intuitive aspects of operationalism at the information level.

At the operational level of the physical encoding of information, since it's actually about operation on matter and energy therefore, information, understood in operational terms, possesses the same objectivity as matter and energy. This duality constitutes both the objectivity of understanding information and the basis for understanding its subjectivity and the interrelation of the two. On one hand, this is the advantage of understanding information at the operational level, but on the other hand, it is also its limitation. If information is considered to have the same objectivity as matter, our understanding of the relationship between information and matter is inherently limited. In fact, it is the difference in objectivity from matter that fundamentally distinguishes information from matter and energy. This further deepens our understanding of the significance of the specificity of the PIIMO.

It's precisely because of the specificity of the PIIMO, we can see two subtly intertwined aspects. On one hand, the PIIMO is undoubtedly significant, as it not only makes information operations possible due to this identity but also provides a basis and reliability for us to confirm that these physical operations are indeed information operations. It is through the identity between information and matter/energy in actual operational terms that we can gain a deeper understanding of the intrinsic connection between information and energy. On the other hand, the PIIMO does not imply that we can equate information and matter outside the context of operational meaning. This relates to the root of the misconception of "information conservation" in information understanding. Thus, building upon the PIIMO, we can not only deepen our understanding of issues related to information grounding and quantification but also clarify many ambiguities in the understanding of information and its relationship with matter and energy at a fundamental level, and even transcend the information encoding ceiling in information understanding.

6.4　The PIIMO Understanding of Information Conservation

Deriving the conclusion of information conservation based on the fundamental viewpoint that "information is physical" is undoubtedly a logical necessity. However, information conservation clearly contradicts the nature of information creation and common knowledge. By applying the PIIMO, we can not only clearly see the issues with information conservation at a theoretical level but also understand the rational factors underlying it.

Building upon the PIIMO, it becomes evident that what is discussed about the viewpoint of "information is physical" is not information itself but the secondary sources as physical encoding of information – more and more it is physical encoding regenerated based on digital encoding of information, because the discussions about information processing actually mainly refer to the processing of physical encoding of information. The processing of information encoding is essentially physical processing, and therefore, it must adhere to the laws of physics. Both aspects share an identity, which is manifested in the fact that the energy consumption associated with the processing of information physical encoding has a quantifiable relationship with matter and energy processing itself. This fact can be fully confirmed through authoritative explanations and experimental validations of information conservation: "Shannon showed that sending a bit of information over a telephone line required a minimum amount of energy, roughly the amount of energy possessed by a typical molecule bouncing around at room temperature." In information processing, erasing information always produces heat that escapes into the environment. "The fact that erasing a bit requires a minimum loss of energy is now widely known among computer physicists as Landauer's principle" (Siegfried, 2000, pp. 70, 71, 72). It is clear that what is argued in this way is still the conservation of physical encoding of information rather than conservation of information itself.

From this, we can further see that there are profound cognitive roots in the physical understanding of information. Approaching the PIIMO as an ontological interpretation leads to a series of erroneous conclusions derived from the understanding of the material paradigm of information phenomena. The idea that information, like matter and energy, is conserved is one of the most incredible viewpoints. In the context of information theory, this perspective is not only widespread but also highly natural. Stephen Hawking once proposed a conjecture:

> I think the information probably goes off into another universe. I have not been able to show it yet mathematically, but it seems to be what the physics is telling us. But most other people are rather conservative and don't like the idea of loss of information because it is not what they were taught in graduate school.
>
> (Siegfried, 2000, p. 203)

In 2004, Hawking announced a method to prove the conservation of information: "Indeed, one can regard unitary Hamiltonian evolution about topologically trivial metric as global conservation about quantum flowing through three cycles under global time translation" (Hawking, 2005). Hawking applied a representation of quantum uncertainty to the topology of space-time and concluded that black holes are not entirely black; they also emit information. Hawking wrote, "The confusion and paradox arose because people thought classically in terms of a single topology for space-time." Some physicists found Hawking's new theory ambiguous and left many unresolved questions. However, one thing Hawking made clear: "There is no baby universe branching off, as I once thought, the information remains firmly in our universe. I'm sorry to disappoint science fiction fans." In the end, he lost a bet to John Preskill, a copy of *Total Baseball: The Ultimate Baseball Encyclopedia*, saying, "from which information can be recovered with ease." Hawking said. "But maybe I should have just given him the ashes."(Gleick, 2011, p. 359) In an existentialistic sense, the belief that "information is conserved" is evidently associated with a physical understanding of information, closely related to the viewpoint that "information always exists in our universe." Its underlying cognitive source is the ontological interpretation of the PIIMO. The ontological interpretation of the PIIMO can be traced back to the initial proposition that "information is physical."

The renowned statement "information is physical" was first put forth by the German physicist Rolf Landauer. This proposition is, in fact, the title of his famous paper. Landauer initially presented this proposition primarily in the context of physics, with the intention of providing a physical foundation for information. As a specific application, he stressed that "computation" must be realized in the processing of physical information encoding. Therefore, he proposed a more consistent statement: "Information handling is limited by the laws of physics and the number of parts available in the universe; the laws of physics are, in turn, limited by the range of information processing available" (Landauer, 1991). This symmetric expression of information and matter clearly stems from the fundamental viewpoint that "information is physical." It is this symmetry that exposes the underlying complexities of this proposition. The former of the two symmetric conclusions is easily understood; a vivid example is the calculation of π. "We can calculate π-to any required number of places. But that requires an unlimited memory, unlikely to be available in our real physical universe" (Landauer, 1991). However, the latter is not that simple – how could physical laws be constrained by the scope of information processing? In the sense of realism, in the sense that it is information itself, not just operationally following physical laws, physical laws would be unaffected by information processing, much like an established result that, as a consequence of specific reasons, is not influenced by subsequent reasons that it contributes to. Information can be understood more deeply from the perspective of receptive relation and its status. However, in the context of putting forward this proposition at that time, the conclusion that information itself, rather than the physical encoding of information, follows physical laws based on the PIIMO, is precisely where the misunderstanding lies in the viewpoint that "information is physical."

In Landauer's view, information and matter are in a complex undifferentiated state. He believed that a clock-like model, similar but more modern, led to the perspective that the universe is a computer, causing him to assert that "there is a strong two-way relationship between physics and information handling" (Landauer, 1991). While he claimed not to have gone as far as thinking the universe is a computer, his viewpoint is sufficient to lead to the fundamental understanding that information is physical: "Information is inevitably tied to a physical representation and therefore to restrictions and possibilities related to the laws of physics and the parts available in the universe." From this speculation about the nature of physical laws, it must be concluded that: "The laws of physics are essentially algorithms for calculation. These algorithms are significant only to the extent that they are executable in our real physical world" (Landauer, 1996). In the previous statement about laws, information and matter are chaotic and difficult to distinguish in an ontological sense on the one hand, and show some differentiation in an operational sense on the other. This is most obvious in Landauer's understanding of information:

> Information is not a disembodied abstract entity; it is always tied to a physical representation. It is represented by engraving on a stone tablet, a spin, a charge, a hole in a punched card, a mark on paper, or some other equivalent. This ties the handling of information to all the possibilities and restrictions of our real physical world, its laws of physics and its storehouse of available parts.
>
> (Landauer, 1996)

Obviously, in this context, the "tie" refers to information encoding rather than information itself. The viewpoint that the universe is a computer provides a deeper understanding of the universe and, at the same time, an evident retreat in the understanding of information. This becomes increasingly apparent as Landauer progresses toward the viewpoint that quantum information is physical.

On the basis of an incomplete understanding of information, the connection between information operations and matter almost inevitably leads to their ontological cohesion. In this sense, Landauer's insights are quite remarkable. He almost touched upon the PIIMO even before a comprehensive understanding of information was in place: "Computation is inevitably done with real physical degrees of freedom, obeying the laws of physics, and using parts available in our actual physical universe" (Landauer, 1991). Landauer's viewpoint clearly indicates that the extensive effort he devoted to exploring the physical basis of information pointed toward the carrier of physical encoding of information. Unfortunately, Landauer passed away shortly after publishing this paper, leaving no opportunity for further elaboration. In fact, if we truly inquire about the physical basis of information, it is the physical encoding of information. The viewpoint that "information is physical" directly arises from the fact that information cannot exist but is based on matter and energy. From this, the understanding root to derive the paradoxical conclusion of information conservation from "information is physical," leading to the dilemma, is fully explained.

In summary, all information is created ex nihilo, but only after it undergoes physical encoding does the physical encoding of information become "conserved." Clearly, this sense of conservation still refers to the conservation of physical encoding rather than information itself. It implies not the conservation of information itself but the conservation of the physical encoding of information. The misinterpretation of the conservation of information encoding as the conservation of information itself is objectively rooted in the PIIMO. When this principle has not been unveiled, what can be observed is the direct identity of information and matter in the forms of biological or physical manifestations of information and matter in their encoded forms. Elevating the relationship between information and physical encoding of information to a higher level makes this even clearer.

Bits are not information but digital encoding and its unit of measurement. However, information can only be measured through its physical encoding. Genes are the biological encoding of information, and DNA is the biological "house" of this information encoding, neither of which is information itself. As a process of receptive relation, information only becomes "embodied" (materialized) and takes on a tangible form after being grounded on physical encoding. Information can only be processed and even transmitted when it is grounded in physical encoding, allowing people to manipulate information and engage in information processing and production.

6.5 The Deepening Understanding of Information Grounding and Quantification Issues in the Context of PIIMO

According to the PIIMO, we can further enhance our understanding of important issues related to information grounding and quantification. The actual manipulation of information and the symbol grounding are inherently interconnected, and they are closely related to the encoding, especially the physical encoding of information.

It is precisely through the physical encoding of information that symbols as information encoding can finally ground, allowing information to enter physical operations. It is also in this development process that digital encoding of information has a special status and significance. As a special symbol encoding, what digital encoding of information involves is an ideational issue. One bit of information is the amount of information needed when making a decision between two equally probable choices. If we want to determine whether someone's height is less or greater than six feet, and we know both possibilities are equally likely, then we only need one bit of information. Two bits of information allow us to make a decision among four equally probable choices; three bits allow us to make a decision among eight equally probable choices and so on. In other words, if there are thirty-two equally probable choices, we need to make five binary decisions, each with one bit of information, to determine which choice is correct. Therefore, the basic rule is simple: the number of bits required for information increases each time the number of possible choices doubles. (Miller, 1956) To implement this series of algorithms based on numerical encoding of information, it is necessary to directly convert numerical encoding into physical encoding of information.

Without recognizing this critical subtle conversion and the operation of physical encoding converted from it is regarded as digital encoding of information, it will not only treat digital encoding as information itself but also regard operation of physical encoding as direct operation of information itself. From this, two important facts become very clear: on the one hand, bits are units of numerical encoding, not units of information itself; on the other hand, the so-called quantification of information is actually the quantification of information encoding. What we typically refer to as the quantification of information is, in reality, the quantification of physical encoding of information – often it's digital encoding.

It is through information encoding that the possibility of quantifying information processing based on the PIIMO can be realized. Shannon established mathematical principles about information based on this. Ralph Hartley even proposed a somewhat related formula (Hartley, 1928). It's impossible for us to realize that what is quantified here is not information but information encoding without the receptive relation understanding of information.

About the quantitative nature of signal understanding, Dretske describes:

> Communication theory does not tell us what information is. It ignores questions having to do with the content of signals, what specific information they carry, in order to describe how much information they carry. In this sense Shannon is surely right: the semantic aspects are irrelevant to the engineering problems. Nevertheless, in telling us how much information a signal carries, communication theory imposes constraints on what information a signal can carry. These constraints, in turn, can be used to develop an account of what information a signal does carry. In this sense Weaver is surely right: communication theory is not irrelevant to a specification of a signal's information content.
>
> (Dretske, 1982, p. 41)

When not linked to the content of information, treating the physical encoding of information as information itself poses no problem for information processing, but in doing so, it not only limits the understanding of information to the operational level but also precludes involvement with more complex information content. Based on the PIIMO, not only can the understanding of information quantification be deepened but also research on symbol grounding at the information level can be related to embodiment issues.

The PIIMO means that only through specific encoding can information as receptive relation be measured. Thus, the understanding and quantification of information grounding and quantification can be further deepened. Due to its different levels when information encoding is treated as information itself, the complexity of information encoding makes the concept of information very complex. This is why there has been a long-standing lack of consensus on the definition of information. Currently, although the measurement of information is actually about the measurement of information encoding, the confusion between information and information encoding has objectively caused the complexity of measuring information encoding, resulting in corresponding misunderstandings. Information encoding has different types and levels, or we can say that the carrier of information encoding has different levels. In empirical science, since the research object is matter and energy, treating information encoding as information itself is, in fact, treating the physical encoding of information as information itself, thus limiting the understanding of information and information processing to the empirical scientific field. As a result, understanding of the PIIMO and its physical limits is not clear, and the problem of information grounding and quantification cannot be truly clarified.

Therefore, it is clear from this principle that, in the ultimate sense, information grounding is not ground to symbols or other ideas, but grounding to the physical encoding of information. Currently, information quantification is not about information itself but the quantification of the physical encoding of information. Whether in information grounding or information quantification, the PIIMO holds a fundamental position. Thus, as an undisclosed principle, it involves almost all the origins of misunderstandings about information and its relationship with matter and energy in scientific research. It is crucial to clarify the deep-seated fog in understanding the relationship between information and matter/energy in the understanding of information.

6.6 Conclusions

In the realm of natural sciences, particularly physics, the notion that "information is physical" has long held sway. The reasoning behind it is both complex and simple. On one hand, in the domain of material sciences, not taking this fundamental viewpoint as a premise implies that the study no longer falls within the domain of physics, which focuses on matter and energy. On the other hand, as receptive relation, information must be grounded on matter and energy. In the ultimate sense, information is receptive relation based on matter and energy. Regardless of how one interprets it, if information is understood as physical encoding of information, akin to a telecommunications signal, then information is inherently physical. However, interpreting information itself as matter and energy obviously reduces information to that in the final analysis. It is widely accepted that information cannot be reduced to matter and energy; this is a fundamental principle in information research. As receptive relation, information can only exist and be practically operational when it is embodied in physical encoding. Thus, there exists a significant tension between these two aspects, harboring the potential for theoretical breakthroughs and even specific mechanisms.

Because as receptive relation, information must ultimately land on matter/energy translate into physical encoding to be practically operational, and because the physical encoding of information must adhere to physical laws in operation, a logical rift emerges between the existence of information and its practical operation. Logical rifts often signify higher levels of entry. It is from this point that a special fundamental principle arises concerning the intrinsic fundamental connection between information and matter/energy can be derived: the PIIMO.

As a special principle, the PIIMO holds unique significance. It not only centrally reflects the fundamental relationship between information and matter/energy but also embodies the distinctiveness of information from it. This implies that, in an ontological sense, information is neither matter nor energy; yet in an operational sense, the physical encoding of information in physical operations follows physical laws. Information, in practical operation, must inevitably adhere to physical laws, but the action and effects of physical laws do not have a direct regulatory connection to information in an ontological sense, resulting in a symmetry breakdown in the relationship between information and matter/energy. In the receptive relation understanding of information, the physical encoding of information is not the information itself, but a secondary source of information. In terms of existence, it is an overlap of matter and energy and information. It cannot be viewed as purely existing in the form of matter and energy, and even less so as the pure form of informational existence. It can only be regarded as the physical way in which information exists in an operational sense. This indicates that, due to the necessity for information operations to translate into physical encoding, information and matter/energy share an identity at the operational level. Therefore, the PIIMO has a physical limitation and is only applicable within the scope of information operations.

Although the PIIMO is operational in nature, its significance is far from ordinary. Based on this principle, systematic research not only clarifies many important related understandings, including the dilemma of "information conservation," issues closely related to the symbol grounding, and issues of information measurement and quantification, but also provides a deeper clarification of the boundary between quantum physics and classical physics. Moreover, it can more precisely articulate the mechanistic relationship between information and matter/energy. Building upon the receptive relation understanding of information and delving into the PIIMO, the understanding of the relationship between information and matter/energy can not only delve to the deepest level, exploring the relationship between physical encoding and ideational encoding of information, but also more specifically involve the relationship between the material paradigm and information paradigm, along with the corresponding paradigm shifts.

The relationship between these two fundamental information encodings signifies a higher-level connection between the worlds of information and the material world, while the relationship between the two basic paradigms provides the deepest foundation for furthering our understanding of information in comparison to matter and energy. At the level of the PIIMO, we can better derive the basic characteristics of information in the closest and most fundamental comparison with matter and energy.

7 Basic Characteristics of Information

7.1 Introduction

As a presence distinct from material, information possesses basic characteristics that set it apart from material. Furthermore, just as there is a relationship between the understanding of material and its basic characteristics in the study of it, in the study of information, the understanding of information and the understanding of its basic characteristics form a bidirectional interpretation. The understanding of information determines the understanding of its basic characteristics, and conversely, the understanding of the basic characteristics of information influences the degree to which information phenomena are understood.

In the material paradigm, the understanding of information determines the depth of understanding of its basic characteristics. The most typical is the information encoding understanding of information. If information is understood merely as the physical encoding of information, its basic characteristics are inevitably overshadowed by the characteristics of material, leading to a limited understanding of the basic characteristics of information. On the other hand, interpreting information as the ideational encoding, and building upon the ideational system based on it, it will guide the understanding of its basic characteristics toward ideational understanding, thereby moving away from the empirical nature of information. That's why "the nature of information itself, remains unclear" (Mattingly, 2021, p. 165). Without a thorough understanding of information, it is impossible to systematically derive the basic characteristics of it.

Firstly, whether in the source-carrying understanding or the signal-preexisting understanding of information, information exists objectively like matter and energy, regardless of whether there is an information receiver. From the existence of information in this way, it is impossible to explain or even see the fact that information can be generated by the receiver. The receiver-assigning understanding of information does show the generative nature of information by the receiver, but due to "overcorrection" it has gone to the opposite side. Admittedly, this information understanding itself implies that information is generated by receivers, but this understanding of information not only cannot explain how receivers generate information without an ideational system, but also cannot clearly and sufficiently explain the objectivity of information. Not only is there a lack of differentiation between the source and the receiver, but the informational activities of plants are

DOI: 10.4324/9781003484851-7

entirely objective processes, and in simpler unconscious information activities of animals and even humans, there is no receiver-assigning. This reflects the problem of paradigm shift. As long as information understanding is within the material paradigm, it cannot truly reveal the nature of information creation out of nothing (the created ex nihilo of receptive relation).

Secondly, the source-carrying, the signal-preexisting, and the receiver-assigning understanding of information can only explain phenomena of irreducible emergence in matter and energy at a certain level and cannot truly address deeper levels of irreversible and irreducible emergence. Currently, the study of emergence is mainly concentrated in the field of systems science and emergent theory in condensate physics. In the field of theoretical physics, research on emergence in emergent theory is, as the name suggests, about phenomena of matter and energy; in system theory, emergence refers to new characteristics exhibited at the whole level by physical system that cannot be explained by single elements or subsystems. Due to being completely different from physical mechanisms, emergence in the sense of information and emergence in the sense of matter and energy belong to two different levels. Research on phenomena of emergence in information should surpass the study of emergence in condensed matter physics and system science in the clarification of the level of emergence.

Thirdly, due to the inclination of understanding information at different degrees of realism, the source-carrying, the signal-preexisting, and the receiver-assigning understanding of information cannot reveal the relational nature of information. Consequently, it is impossible to address the corresponding basic characteristics of information. For instance, it is impossible to understand, in the sense of matter and energy, phenomena like this: at least two telephones are needed to constitute a communication information relationship. The more complex the information relationship is, the less likely it is to be understood based on the material paradigm. In the case of advanced stages of human social information activities, understanding the relationships between sources and receivers becomes even more challenging at higher levels of information processes. In the end, without a relation paradigm, it is impossible to truly understand the basic characteristics related to this phenomenon: the interdependence, indivisibility, mutual influence, common action, and mutual stimulation among the constituent parts within the whole process mechanism.

Finally, as the most subtle reflection of the fundamental difference between the material paradigm and the information paradigm, whether it's the source-carrying, the signal- preexisting, or the receiver-assigning understanding of information, all fail to distinguish between the nature of material sharing and informational sharing. In fact, in English, these two concepts can only be expressed with the same term. Without being within the information paradigm, it's impossible to grasp the nature of matter and energy being only shareable in the sense of matter and energy and not in the sense of information, or to differentiate the fundamental difference between material sharing and informational sharing, thus failing to reveal the nature of informational sharing that goes beyond material sharing in the context of information. Without the receptive relation understanding of information, it's impossible to fully understand the nature of informational sharing.

From this perspective, the basic characteristics of information can only be systematically derived based on a paradigm shift. Due to the involvement of paradigm shifts in information, the receptive relation understanding of information provides new conditions for research on basic characteristics of information and lays a premise-based foundation for systematic research on basic characteristics of information. Based on the receptive relation understanding of information, basic characteristics of information can be systematically derived: created ex nihilo, emergence, reciprocity and sharability.

7.2 Created Ex Nihilo of Information

As receptive relation, the relational nature of information is fundamental. Indeed, there are relations in matter and energy without receptivity, but those relations are physical and chemical interactions between material entities and fields, which do not have effects beyond physical interactions, let alone relation between concepts. As receptive relation, information has non-entity properties. Non-entity means that it does not have the extension possessed by objects, which is the most prominent feature distinguishing information from matter that extends beyond the physical interaction. This is especially distinct from the relational nature of information, which possesses a non-material quality. Non-material implies the absence of the extension that physical objects have, and this is the most prominent feature that distinguishes information from matter.

Unlike material entities with extension, information lacks extension. The fundamental difference in extension between information and matter stems from the substantiality of matter and the relational nature of information – and this is receptive relation, as specific processes. This relational existence is entirely different from the existence of matter, which cannot be created or destroyed. From an ontological perspective, information is created based on matter. This characteristic stands in direct contrast to the conservation principle of material entities and fully demonstrates the non-conservative nature of information.

The fact that receptive relation is created out nothing implies that information is a creation ex nihilo. It is from the understanding of the receptive relation of information that we can clearly define the first basic characteristic of information: created ex nihilo. The essential nature of information is crucial, and it cannot simply be understood as "creativity," because, to be precise, all information is created based on matter and energy. Therefore, the created ex nihilo of information implies its non-conservative nature. In contrast to the conservation of matter and energy, information possesses a non-conservative nature, which serves as an ontological proof of the created ex nihilo of information.

Material with extension cannot be destroyed or created; whereas, as receptive relation, information is not an established inherent existence like matter and energy. Instead, it is initially generated through the interaction of the organismic receptivity. After differentiation between receiver and source, it further develops into the typical form generated through the interactions of sensory receptivity. As an effect of receptive interaction, the very nature of receptive relation implies that information is created out of nothing, and created ex nihilo is an inherent characteristic of receptive relation. As receptive relation, typical information essentially means it is a relational process created by the receiver through receptivity. Information is the effect generated by the direct or indirect interaction between the receiver and the source.

As an effect process of receptive interaction, information is not only created out nothing, but it ceases to exist once the receptive activity ends. For instance, quantum phenomena are generated by observations as a receptive activity. When the observation ceases, the quantum phenomena simultaneously disappear. Therefore, information cannot be conserved like material. Due to the non-conservation of information, the creative role of the agent (receiver factor) is prominently highlighted. This is precisely the unique nature that sets information resources apart from material resources. A profound understanding of the created ex nihilo of information holds immeasurable significance not only for human existence and development but also undoubtedly contributes to a deeper understanding of human nature.

As intelligent beings embodied in living organisms, the most fundamental basis of human existence is matter and energy. Since matter and energy cannot be destroyed or created, the created ex nihilo of information holds special significance for the unfolding of human nature: creative activities are the most in line with human nature, and creative activities directed toward matter and energy can only be transformative with creativity, while creative activities in the sense of information are fundamentally creative in the truest sense.

Understanding information through this fundamental characteristic of created ex nihilo not only reveals many new questions at the informational level but also presents new aspects and understanding approaches to many existing questions. This not only provides us with a deeper understanding of the world and its relationship with humanity but also allows us to have a more profound grasp of human knowledge, especially contemporary scientific knowledge.

In the fields of physics and biology where information content is involved, information is often considered to possess conservation, similar to energy. Due to the inability of the peculiar quantum phenomena to be explained by material principles, quantum physics is more keen on research on information within the material paradigm; because research objects for quantum physics are still matter and energy in quantum information research they even treat information as being like matter and energy thus there came about principles of conservation of information. Richard Jozsa argues, "a notion of quantum information is based on the presence of entanglement in (multipartite) quantum systems, leading to an information conservation principle (corresponding to the fact that in a closed system, entanglement cannot be changed by local operations)" (Jozsa, 2004). The conservation of information obviously conflicts fundamentally with the created ex nihilo of information. The paradoxical nature within this, in fact, indicates that the source-pre-existing understanding of information ultimately leads to an understanding rooted in matter and energy, and even suggests that an understanding of information in a substantial sense will eventually lead to self-contradictory conclusions. It is only because physics inherently focuses on matter and energy as its object that the inherent dilemma of information is "bluntly" obscured.

Based on the relationship between receiver and source, the created ex nihilo of information gives it a very rich and distinctly different nature. The creation nature of information is not only related to the characteristics of the receiver and source but also closely tied to the nature of their receptive interaction. The weaker the information agent as receiver is, the stronger the effect of the source, resulting in a stronger objectivity of the created information, to the point of being primarily an objective representation, such as when the source (object) is a natural existence. In the organismic receptive relation in which there is not yet differentiated between the receiver and source, the created information possesses a typical objectivity. The stronger the information agent as receiver, the weaker the effect of the source, resulting in a stronger subjectivity of the created information, and it can even be considered subjectively generated, such as when the source corresponds to a non-existent specific individual existence. The former is typical in the natural sciences, while the latter is typical in literature and the arts.

The created ex nihilo of information is an undisputed basic fact in both literature and the natural sciences. As people naturally express, "information is continually being created in social life as we try to make sense of our worlds" (Lynda & Paul, 1991, p. viii). From this, based on the understanding of receptive relation of information, we can further see the epoch-making implications of the created ex nihilo of information. It clarifies the crucial characteristic that distinguishes information resources from material resources: relative to the material resources, the created ex nihilo of information implies that information resources are boundless. Humans are the highest-level receivers, and the boundlessness of information resources holds extremely rich and important significance for humanity. Due to conservation, matter and energy cannot be created, and material resources are finite. Due to non-conservation, information creating out of nothing and information resources are infinite. Not to mention bar code resources, even Arabic numeric encoding is boundless. The Indian story of arranging grains in a chessboard grid in a geometric progression is an illustrative example of the contrasting situations between information resources and material sources.

The created ex nihilo nature of information implies its richness in resources, even associating with its value characteristics at a deeper level. This allows us to perceive a deeper understanding of the relationship between facts and values. On one hand, in the early stages of information development in a receiver, the facts and values of information as receptivity interaction effects tend to be chaotic and undifferentiated. As the receiver develops, the hierarchical ideational system based on the ideational encoding of information becomes higher, encompassing more systematic value factors. Receptive relation established within the holistic guan zhao of the ideational system becomes more profound. If we consider the ideational system as the embodiment of truth, information facts and values will be completely differentiated. However, if we understand the status and role of the ideational system as a holistic guan zhao for establishing further receptive relation, information facts and values remain integrated.

On the other hand, all information is generative, but only consciously aware information generation is truly creative. It is through information that not only does the subjectivity of humans as the highest-level receivers truly manifest, emphasizing their subjective status, but also highlights the creativity and value of the receiver.

The created ex nihilo receptive relation also implies that information possesses a compositional nature completely distinct from matter and energy. As receptive relation, information has a mechanism of composition entirely different from that of matter and energy. The more complex the information activity, the more intricate the involved mechanisms, and the more the created ex nihilo of information manifests itself as a typical emergent process.

7.3 Emergence of Information

The receptive relation nature of information not only implies created ex nihilo but also emergence. Regarding information as receptive relation, the basic mechanism of emergence is the same as that of created ex nihilo, but the specific mechanisms of emergence and created ex nihilo are entirely different.

The emergence of information refers to the irreducibility of information generated within the receptive relation. This irreducibility exists at a more fundamental level and holds a more essential significance. From the emergence of information, two levels of emergence can be observed: material emergence and informational emergence. It is through these corresponding levels of emergence that two levels of irreducibility are correspondingly demonstrated.

Irreducibility is a common feature of emergent phenomena, but it exists at different levels. Two fundamental levels are the irreducibility of composition or constituents and the irreducibility of attributes or functions. The material emergence and its irreducibility are in the sense of constitution, while the informational emergence and its irreducibility are in the existential sense. Differentiating between emergence in biological systems and consciousness, we can see this level of distinction. "It has been argued that meaning emerges in the domain of biology, while consciousness has been held to be an emergent property of integrative levels of information" (Bawden & Robinson, 2013). Emergence in biological systems can be a manifestation of material emergence or informational emergence, whereas consciousness is purely an informational emergent phenomenon. The irreducibility of attributes or functions refers to the fact that the new qualitative attributes or functions possessed by the emergent physical structure cannot be reduced to the sum of attributes or functions of its constituent parts, but the physical constituents or components are reducible. For instance, when water turns into ice, the attributes or functions of ice are irreducible, but its constituents or components are reducible.

Research on the irreducibility of attributes or functions began early in science and philosophy. As an exceedingly complex phenomenon, this form of emergence has been extensively studied, particularly in the field of contemporary sciences such as complex systems science and condensate physics. Since information phenomena were not clearly distinguished from physical phenomena in the material paradigm, there have been various perspectives on the study of emergence, making it difficult to reach a basic consensus. Moreover, the study of emergent phenomena in this paradigm only focused on the emergence itself, without delving into its underlying mechanisms, inevitably leaving critical aspects unexplained. Discussing emergent phenomena in the material paradigm does not allow for a deep understanding of the information mechanisms that serve as the origin of emergence. The receptive relation understanding of information provides a new foundational basis, allowing for a deeper understanding and grasp of emergence, especially in complexity science, particularly as a contemporary branch of theoretical physics known as condensate physics, where emergence is universally present.

The emergence of physical properties primarily refers to the appearance of higher-level structures in physical systems, resulting in the emergence of new qualities or functionalities that were previously unattainable and irreducible. On the other hand, the emergence of information is a characteristic inherent in the formation of receptive relation and is distinct from the formation of physical properties. In the case of information, the receptive relation itself emerges.

In contrast, the emergence of information has completely different characteristics. Even the most fundamental receptive relation does not form gradually; instead, it emerges, and the entire emergence process is total irreversible and irreducible. In the context of receptive relation, the emergent nature of information lacks reversibility and reducibility in terms of components or composition.

The more complex the information activity, the more information generation manifests as emergence. In increasingly complex information phenomena, the level of emergence is higher and more typical, only because informational emergence relies on physical processes, it can be challenging to recognize this fundamental point amidst various material emergence phenomena: compared to the emergence of information, matter and energy do not possess a fully meaningful sense of irreversibility and emergence. The emergent properties and functionalities that arise in the evolution of systems on new qualitative levels and their structures and functions are, in fact, manifestations of the emergence of information. The phenomena studied in condensate physics, for instance, are fundamentally different from reductionist physics; they are manifestations of the emergence of information. However, these studies still fall within the traditional domain of physics. The physical manifestations that represent the emergence of information cannot be traced back to their information mechanisms. Just as consciousness, the emergent properties of brain neural systems are manifestations of the emergence of information. Even when individuals form new qualitative structures or functionalities as part of a group, it is fundamentally due to the emergence of information. Because the emergence of information is an inherent characteristic of receptive relation, and receptive relation itself exhibits irreversible emergence, it is impossible to reverse-engineer the information mechanism from the physical carriers of information agents.

Since the emergence of information must be implemented through physical processes, there has been research on emergence phenomena in the field of physical systems for some time. With the development of information research, people have become increasingly aware of the connection between emergence and information. David Bawden and Lyn Robinson, for instance, observed,

> It is clear, when we examine the subject matter of the library and information sciences that there are very real and important concepts which are not captured in the objective view of physical information. They include, but are certainly not limited to, knowledge, meaning, understanding and relevance. These are emergent properties, seen at the level of communicable human information, but not at the physical level.
>
> (Bawden & Robinson, 2013)

The higher the level of information development, the more apparent the connection between emergence and information becomes. With the development of the receiver thereupon then receptive relation, the richness and complexity of information emergence increase.

The richness of information emergence is most evident in the emergence of meaning. An elevation in the holistic level implies an upgrade in meaning. However, this understanding primarily arises from intuition about the relationship between information and emergence and does not yet represent an understanding of emergence at the fundamental level of the basic characteristics of information. It is precisely because of this that it is impossible to connect the emergence of information with its physical (hierarchical) manifestations. Some argue that "emergent properties do not form a hierarchy, with the lower levels being more fundamental" (Bawden & Robinson, 2013). Only by the receptive relation understanding of information can we see that, as a fundamental characteristic of information, emergence indeed has a hierarchical structure. However, no matter how rich and complex the hierarchical structure may be, as a fundamental unit, the internal mechanisms of any hierarchical structure are irreversible and irreducible. Emergence does not only exist at these two fundamental levels but also encompasses rich content in between them.

In terms of emergence in a general sense, there are two different connotations of "emergence": the material emergence and the information emergence. Through an understanding of receptive relation of information, it becomes clear that information emergence is the emergence in the real sense, while the material emergence is more accurately described as mutation. Moreover, in the process of information evolution, the corresponding material emergence is essentially the material manifestation of information emergence, exemplified typically in the mutation of biological organs such as the eye during the process of biological evolution. It is in this sense that emergence is a fundamental characteristic distinguishing information from energy. The relationship between these two different connotations of emergence can be clearly observed in a particular emergent phenomenon. An important form of emergence is collective emergence which lies between physical and informational emergence. As a specific manifestation of informational emergence, collective emergence represents a direct physical manifestation of informational emergence mechanisms, making it particularly useful for understanding the relationship between these two fundamental forms of emergence. Therefore, in the study of emergence phenomena, emergence at the level of individuals to groups holds special significance.

Emergence is not only evident at the individual level within information agents but also manifests at the level of groups, particularly in the field of systems theory, within systems groups. A systems group consists of individual systems with specific homogeneous functions, and its behavior exhibits certain collective properties. On one hand, the behavior of a systems group is the collective result of the behaviors of all individual systems within it. On the other hand, the behavior of individual systems within a systems group is to some extent influenced by the collective behavior.

Researchers like John Wheeler and later Kevin Keller have conducted extensive studies in this area. Keller noted that "Wheeler saw 'emergent properties' within the superorganism superseding the resident properties of the collective ants. Wheeler said the superorganism of the hive 'emerges' from the mass of ordinary insect organism" (Kelly, 1995, p. 14). In the context of collective emergence, the distinction and connection between informational emergence and material emergence are most typical: the informational mechanisms of emergence cannot completely be reduced, but material emergence, as a result of informational emergence, is reducible in terms of attributes or functions, although it may not return to its original state once reduced. The emergence of individuals forming a group is reducible, although the new quality does not exist after reduction; only the intrinsic emergence of informational mechanisms is truly irreducible, meaning it cannot be reduced to a state prior to emergence.

The reducibility of the emergent products of individuals forming a group is related to material emergence, while the irreducibility caused by informational emergence is what is truly irreducible. This is why informational emergence not only relates to the formation of groups but also to the formation and development of wholes. Therefore, when delving into the study of emergence at the informational level, it touches upon issues of holism and reductionism at a deeper level.

Understanding the relationship between holism and reductionism at the level of informational emergence can lead to fundamental progress in two aspects. On one hand, the emergence of information provides new fundamental facts for a deeper understanding of holism and establishes specific conditions for the effective scope of reductionism. On the other hand, based on receptive relation understanding of information on the basis of big data constituted by digital encoding information can lead to a deeper understanding of the relationship between reductionism and holism.

In the field of information, emergence is not only a common phenomenon but also signifies the important nature of information. The study of information as receptive relation is becoming increasingly important in the study of emergence. Deepening the receptive relation understanding of information can further clarify its important nature due to the distinct emergence in contrast to material mutation: the irreversibility different from that of matter and energy.

As receptive relation, the emergence of information implies the irreversibility of information relational processes of information. The irreversibility of information evolution is a result of the irreversibility of receptive interactions. This is closely related not only to receptive relation itself but also to the evolution of information. In biological evolution, we can observe typical manifestations of information evolution. "We have seen that organic information is a non-reducible entity, because it cannot be described by anything simpler than its sequence, and the same is true for organic meaning, which cannot be defined by anything simpler than its coding rules." (Barbieri, 2016) From the nature of information, it becomes clearer that the reason biological or physical reduction of information is impossible lies in the fact that information cannot be reduced to matter and energy. In other words, there is a fundamental boundary of physical reductionism for information. All matter and energy can, in principle, be reduced to physics or chemistry, but information agents cannot. This is because information and matter/energy belong to completely different existences possessing completely different properties – truly emergent nature in a complete sense of characteristics, characterized by true irreversibility.

The relationship between the reducibility of physical mechanisms at the level of components or constituents and the irreducibility of information mechanisms can be understood as follows: as biological beings, the brain and nervous system are, in principle, reducible at a physical level. However, consciousness, as an information agent, is fundamentally irreducible. This is why even if we fully understand the biological structure of the brain; we cannot comprehend its information mechanisms. In fact, without a complete grasp of the information mechanisms underlying consciousness, a thorough understanding of consciousness is impossible, even when the brain's structure is fully understood. We can only interpret information mechanisms by delving into information as receptive relation and by leveraging the physical manifestations of information mechanisms. This deepens our understanding of the relationship between explanation and interpretation on the informational level: they form a bidirectional cycle. This dilemma faced by consciousness and life researches can find a path in the exploration of artificial intelligence, especially in its quest for generalization.

Regarding the emergence as physical phenomena, we can explain them through the analysis of internal relation, i.e., structure or components. However, when it comes to the emergence of information, we can only interpret it through its external relation, i.e., function or behavior. When explanation and interpretation form a bidirectional cycle, a complete mechanism of understanding is established. Due to the differences between the emergence of information and physical phenomena, it is only by unveiling the nature of information emergence that we can truly understand the bidirectional cycle of explanation and interpretation. This is particularly evident in solving the mystery of consciousness, where neuroscience addresses the "easy problems" of consciousness through explanation, while information research addresses the "hard problems" (Chalmers, 2010, p. 4) through interpretation. The latter involves the information mechanisms underlying the emergence of consciousness. Whether it's the emergence of meaning or structure, both involve an intrinsic connection between information as receptive relation and physical mechanisms. Because the brain is not only a holistic system but also cannot be studied through anatomical dissection in living subjects, brain science faces several limitations.

The evolution of information is simultaneously an unfolding process of information, which has reached its current stage with the development of information technology, symbolized by big data and artificial intelligence. The emergence of information indicates that the higher-level an information agent is, the more it contains emergent higher-level properties, and its mechanisms are less likely to be reducible through its physical carriers. This means that the emergent mechanisms of information are characterized by symmetric irreversibility. The higher the quality of an information system, the less likely it can be understood solely through the exam of its physical carriers, and it must be understood through an understanding of its information mechanisms in order to establish or rebuild the information agent. This implies that it is not only impossible to decipher the general intelligence mechanisms of the human brain solely through physiological anatomical analysis but also suggests that the potential for imitation in secondary innovation of information products decreases as the level of information products increases. This is particularly relevant for research paths in the field of artificial general intelligence and the importance of original basic theoretical research. The emergence of information and the resulting irreversibility of information systems have important implications, undoubtedly, for the social development of the information civilization era.

As receptive relation, the higher the level of the basic characteristics of information, the more directly they relate to human social development. Not only does the nature of receptive relation as information constitute emergence with group and even social significance but also forms more fundamental significance of reciprocity within humanity.

7.4 Reciprocity of Information

The advancement of information technology, particularly in the realms of big data and artificial intelligence, has allowed for a more profound exploration of the essence of information. Among these advancements, one stands out as fundamental and crucial: the exploration of reciprocity in information.

The receptive relation understanding of information elucidates a critical aspect of its essence. Unlike matter and energy, the most fundamental attribute of information is reciprocity. As receptive relation, information evolves from the organismic receptive relation to the sensory receptive relation. This development, from undifferentiated to differentiation between source and receiver, highlights the concept of reciprocity. Since humans are the products of the evolution of information to an advanced stage, reciprocity is not only a concentrated manifestation of the receptive relation nature of information but also a fundamental characteristic of anthropology. Therefore, for humans, it represents the most vital aspect of information that involves a fundamental paradigm shift.

Reciprocity expresses the interdependence, indivisibility, mutual influence, common action, and mutual stimulation among components that constitute a whole in the holistic process mechanism. The higher the level of information relationships, the more evident reciprocity becomes as a mutually generative process. In essence, reciprocity is the most fundamental characteristic of receptive relation. Among the essential aspects of information, one of the most crucial is its reciprocity.

Regarding the reciprocity of information, the simplest example can be seen in a telephone terminal, which can only function when another telephone terminal is present. On the other end of the spectrum, the most intricate example involves a thinker who cannot exist without the presence of other thinkers. It is in this very sense that reciprocity is not only a fundamental characteristic of information but also an essential feature of anthropology.

Due to an incomplete understanding of information, the understanding of reciprocity has often remained obscured throughout human history. People's comprehension of reciprocity can only be seen when observing higher-level information agents. Evolutionary biology research indicates that the organic nature of groups stems from reciprocity. "An insect colony was not merely the analog of an organism; it is indeed an organism, in every important and scientific sense of the word" (Kelly, 1995, p. 11). However, this type of reciprocity within organisms remains at a lower biological level. As information agents rise to higher levels, individuality becomes stronger, resulting in a weakening of reciprocity among individuals. This is exemplified in the transition from reciprocity to "mutual benefit" (a situation in which two parties provide the same help or advantages to each other). Only by delving into information at its most fundamental level can the true nature of reciprocity be clearly discerned.

Regarding the term "reciprocity," there are different levels of understanding. This concept is often interpreted more as mutual benefit. However, when viewed from the fundamental levels of information and anthropology, it can be comprehended at its deepest level.

The reason why reciprocity is the most fundamental characteristic of information lies in the nature of information as receptive relation. Information represents the interactive process between the source and the receiver, and the receptive relation themselves exhibit correlations from interdependence to mutual stimulation among elements. Receptive relation signifies reciprocity, and it is the receptive relation nature of information that reciprocity develops information agents' progress. The higher the level of information agent development, the stronger and more complex the reciprocity becomes. This has immeasurable and profound implications for deepening our understanding of humanity and human society.

As receptive relation, information not only provides a prerequisite for reciprocity and interpretation but also serves as the ontological foundation for a deeper understanding of information and the phenomena generated based on it, building upon the foundation of reciprocity. As a fundamental characteristic of information and anthropology, reciprocity not only pertains to human sociability and kind characteristics but also offers a deeper explanation for the commonality in humankind needs and their generation and satisfaction.

Reciprocity of information not only reveals a deeper foundation for sociability but also unveils the information needs that underlie human sociability and its development. The sociability of information needs is inherently expressed through the nature of reciprocity of information, without the need for normative explanations based on human self-awareness. In fact, the inherent reciprocity of information is the manifestation of the laws of sociality governing information and information needs. Human information needs can range from instinctual needs that are not consciously perceived to the information needs of the entire human race, sometimes unrelated to individual subjective desires. As a fundamental characteristic of information, reciprocity provides an indispensable theoretical foundation for the in-depth study of human information needs.

As a higher-level information agent carried by living organisms, human beings typify the qualities that are distinct from material needs and information needs. It is the difference between the nature of information needs and material needs that determines the distinct nature of relationships between individuals and society based on material foundations versus those based on information foundations. The relationships between society and individual based on material needs can be contradictory, whereas relationships based on information needs not only won't contradict but can also exhibit commonality: the common needs generated together and their common satisfaction. It is at the level of information needs that we can gain a deeper understanding of the relationship between information, big data, and artificial intelligence. Such research delves into a deeper understanding of human self-awareness.

Through correlations, big data further elucidates the reciprocity of information. Without understanding the reciprocity of information, it is not only impossible to comprehensively grasp the relationships within big data but also to understand the value of these relationships for humanity. As a fundamental characteristic of information and anthropology, reciprocity implies the connection between human needs and the intrinsic relationship of big data, revealing the constructive value of big data based on human needs: starting from human needs and aiming for their satisfaction. Artificial intelligence, on the other hand, expands the reciprocity of information through intelligent behavior, not only facilitating the understanding of the reciprocity between information and big data but also offering a "key" of holistic guan zhao on the relationship between this reciprocity and human values. In the development of artificial intelligence, the superimposition of reciprocity as a fundamental characteristic of information and anthropology is a research topic of great significance. This research delves into a deeper understanding of human society at a more profound level.

As a receptive interaction process between information receivers and sources, information and its reciprocity are most conducive to explaining the relationship between thought and existence, spirit and matter, humans and objects, humans and the world, and the relationship among humans. The receptive relation understanding of information not only further clarifies the most fundamental characteristic of information – reciprocity – but also facilitates a deeper understanding of the ways in which human existence and development are informed.

As the most fundamental characteristic of information, the understanding of reciprocity is key to comprehending other fundamental characteristics of information. It is based on the reciprocity of information that we can achieve a deeper understanding of the shareability of information.

7.5 Sharability of Information

The receptive relation understanding of information can further clarify another important fundamental characteristic of information: sharability. The sharability of information is a fundamental characteristic that unfolds from organismic receptivity to sensory receptivity and gradually becomes typical in the differentiation process between information receivers and sources. As a fundamental feature of information, sharability even involves a hierarchical deepening of the concept of "sharing" as it relates physical properties to information properties.

Due to the lack of understanding in terms of reciprocity understanding of information, the concept of sharing is quite broad, and the true meaning of sharability has not been rigorously defined. Sharing a cake among several people is considered sharing; bicycles and cars placed in public spaces for everyone to use are called shared bikes and shared cars. However, when it comes to the basic characteristic of information, the concept of sharing in an information context is fundamentally different, the difference in nature of "share" between information and matter/energy, which can be distinguished using the concepts of "分享"(fen xiang) and "共享"(gong xiang) in Chinese, needs to be expressed using newly coined words in English, for instance, "di-share" and "co-share," respectively. Connected to the sharability of information, these two concepts can be delineated more clearly. "Di-sharing" is actually sharing in the physical sense, while "co-sharing" is true sharing in the sense of information sharing. In strict terms, the examples of sharing a cake, bicycles, and cars mentioned earlier are all instances of di-share rather than co-share. Sharing information is different essentially from sharing matter and energy in the way that sharing a movie is different essentially from sharing a cake. Matter and energy can only be di-shared, while only information has the potential for co-sharing.

"Di-share" involves jointly possessing an object, indivisibly sharing an object as a whole, separately sharing a part of the same object, or separately enjoying the same object during a process in a diachronic way. In the context of food as material existence, sharing a cake among many people is "di-sharing," and shared bikes and shared cars are more accurately described as "di-sharing bikes" and "di-sharing cars." While sharing in the genuine sense could be in a synchronic way at any time, co-share entails equal ownership of an object, without dividing the object in space and time due to common ownership, and without affecting each individual's complete ownership of the object. As a symbol of celebrating a birthday, having a cake surrounded by loved ones is "co-sharing," even if it occurs in different times and spaces, but can only be "di-sharing" in the material way in the sense of eating it. Similarly, a piece of music, a movie, or the internet being experienced and used by different people at different times and locations is co-sharing.

Thus, the basis for the sharability of information is the receptive relation nature of information: all potential information receivers can establish similar receptive relation with the same (potential) information source without affecting (in terms of information not information source as physical element) the information source itself. The sharability of information not only makes it possess sharability but also makes di-sharing more effective. It is because of this, while the sharability of information is to be highlighted, at the same time it causes confusion between co-sharing and di-sharing. Particularly, information applications can not only improve the efficiency of di-sharing but also even make previously non-sharable products di-shareable, which is why physically shareable objects are named "shared bikes" and "shared cars." In fact, this implies that, thanks to information applications, information's inherent sharability enables physical sharing to be more effective, demonstrating the manifestation and expansion of information's sharability.

The shareability of information is constantly unfolding with the development of information technology, the internet being a prime example of how those secondary sources as physical encoding of information are showcased in an informational and simultaneous thoroughly sharing way.

The inherent nature of information sharing is based on the receptive relation mechanism of information. In information sharing, as long as it is not constrained by physical conditions, all potential receivers can establish corresponding receptive relation with the same information source. Contemporary information technology demonstrates this point vividly; on the internet, anyone can establish a corresponding receptive relation with the same information source within the same server, thereby sharing the same music. This information source is the physical encoding of the music, which serves as the information resource. Receivers can rebuild certain receptive relation with this secondary source of information, allowing the source to be reused repeatedly. The more people listen to it, the more efficiently the information encoding is used, and it does not reduce each individual's share as is the case with physical product sharing. This highlights a deeper implication: the value of information is entirely different from the value of physical goods.

Since physical goods cannot be co-shared, the same physical product, despite having different values due to varying costs, still holds a certain value. However, the value characteristics of information products are entirely different. For information products with the same functionality, if one of them significantly surpasses others in functionality, then in an open environment the other products will remain unused due to lack of opportunity. Large language models are a typical example; if we consider the evolution of models in user usage, the number of users also determines the speed and level of model evolution, creating a gap that becomes insurmountable. If one information product delivers a substantial blow to another in terms of functionality, the affected product not only loses all its value but may even incur negative value – a cost to deal with. This is all due to the inherent co-sharing nature of information.

The co-sharing nature of information arises from the fact that the same information source can be used to establish specific receptive relation with an infinite number of receivers, enabling them to share information accordingly. In terms of specific mechanisms, information can be produced as secondary sources of information through the physical encoding of information. Specific receivers can then rebuild specific receptive relation with these secondary sources of information, allowing for co-sharing of information. Hence, from the co-sharing nature of information, we can see its importance not only for understanding information, big data, and artificial intelligence as a whole but also for human development.

Because of the characteristics of information resources, the co-sharing nature of information is also a basic characteristic with resource property. As a resource characteristic, the co-sharing nature of information is not only significant for humans but also has more speciality. Because information possesses co-shareability that matter and energy don't have, if people are only accustomed to dealing with physical goods, information, with its seemingly paradoxical properties, can be challenging to understand. Ronald Stamper recognized this long ago when he stated, "Information is a paradoxical resource: you can't eat it, you can't live in it, you can't travel about in it, but a lot of people want it" (Stamper, 1985). Paul Beynon-Davies delved deeper into this and pointed out that although information is critical "stuff" it is extremely difficult "stuff" to pin down; it is probably not even "stuff" at all.

> If information is a commodity it is a very strange commodity. If somebody sells information the commodity does not pass from seller to buyer like a traditional commodity such as food; the seller still retains the information. The "consumption" of information is therefore radically different from the consumption of physical commodities such as food, wine and electronic goods.
>
> (Beynon-Davies, 2011, p. 176)

This can indeed be one of the challenges when trying to understand information, but when viewed in the context of the receptive relation understanding of information, it becomes a particularly important characteristic. Based on this, it not only helps us better understand the co-sharing nature of information but also facilitates a deeper and more comprehensive understanding of the reciprocity, created ex nihilo, and holistic characteristics of information.

Due to the receptive relation nature of information, the co-sharing nature of information is related to the inherent nature of information itself. Therefore, information resources not only have the inherent nature of being shareable but also, due to the infinite nature of information resources, can be shared infinitely. For the co-sharing nature of information, the reciprocity of information serves as its foundational characteristic, and it is precisely this interactivity that signifies that co-sharing is the essence of information.

The co-sharing nature of information is inherently related to other fundamental characteristics of information, and within this inherent connection, there are still some issues that need further exploration. For instance, while information has an inherent co-sharing nature, due to the emergent nature of information, information mechanisms possess irreversibility and irreducibility, thus imposing natural limitations on co-sharing. Therefore, when it comes to the fundamental characteristics of information, whether at a local or global level, there is a wealth of informational significance.

7.6 Conclusions

The relationship between information and its basic characteristics indicates that a comprehensive understanding of receptive relation is essential to systematically derive the basic characteristics of information such as created ex nihilo, emergence, reciprocity, and sharability.

As receptive relation, information inherently implies created ex nihilo, and created ex nihilo is a natural characteristic of receptive relation. Clarifying the created ex nihilo of information is conducive to a deeper understanding of human nature and the nature of human knowledge, especially contemporary scientific knowledge. Created ex nihilo is an ontological characteristic of information; it has a completely new ontological implication that can truly cover the concept of "non-existence." This fundamental characteristic has special significance for the unfolding of human nature. Creative activities are the most aligned with human nature, and the creation ex nihilo is the essence of to be created in a thorough sense.

The receptive relation understanding of information presents the emergence in the sense of information, leading to two different kinds of emergence: material emergence and informational emergence. What is irreducible is new quality functions or attributes and its components or composition are, in principle, reducible. However, in informational emergence, there is no reducibility in any sense. The receptive relation nature of information implies an irreducible emergent nature in its entirety. Physical mechanisms are always, in principle, reversible and reducible, whereas only information mechanisms with the integrity of receptive relation possess a genuinely irreducible nature. Genuine emergence is based on the constitutive nature of information based on the creation of receptive relation, possessing an irreducible compositional nature that physicality lacks. It is in this regard, information emergence is the emergence in the real sense, while the material emergence is more accurately described as mutation. The phenomenon of emergence of information, as receptive relation, and its reciprocity are intrinsically linked, with the former being the result of the elevated development of the latter.

Reciprocity is the radical characteristic of information, and the emergent nature of information is its realization. As a process of upgrading, emergence becomes increasingly complex with the development of information. Reciprocity is the most basic characteristic of information stemming from receptive relation, and it constitutes a completely new relational property absent in physicality. It forms the foundation of fundamental characteristics of anthropology. At its most basic level, reciprocity is both the fundamental characteristic of information and that of anthropology. It is reciprocity of information that not only implies the important nature of information needs differ from physical needs: commonality of satisfaction even generation, but also constitutes the information co-sharing property, which physicality lacks.

Shareability is the utilization characteristic of information based on its relational nature and unique reciprocity. Since matter and energy cannot create and lack reciprocity, they can only be di-shared and not co-shared. As receptive relation, information implies not only that different receivers can simultaneously establish similar receptive relation with the same source without the need for source division, thus achieving co-sharing, but also that it can make physical di-sharing more effective or make matter and energy in a state of impossible di-share become possible. It is in this regard, the true sense of sharing lies in the sharability of information. From this perspective, on the foundation of human material civilization, the basic characteristics of information and their integration signify what human information civilization development truly entails.

The basic characteristics of information and its holistic composition and integration effects have milestone significance for the development of human society, especially in the context of the information civilization. Due to the conservation of matter and energy, we can only carry out transformative creation with respect to it. However, information possesses characteristics of created ex nihilo, allowing us to create information out of nothing. Reciprocity is not only a basic characteristic of information but also a basic characteristic of anthropology. Information not only highlights the kind feature of human beings but also, through the development of artificial intelligence, reveals the evolutionary process of information: starting from the simplest receptive relation based on matter and energy, and gradually forming the information agent through the integration of receivers and sources of information, eventually evolving into intelligent agents. With a higher-level holistic guan zhao of artificial intelligence, we can gain a more accurate understanding of the relational nature of information agents and intelligent agents, as well as further clarify the complex usage relationships between humans, information, and matter/energy.

The reciprocity and shareability of information not only signify new levels of expansion but also the commonality of need. This commonality is prominently reflected in the processes of satisfaction even the germination of information needs.

In the development of human information civilization, reciprocity of information is a crucial link. As a basic characteristic of information, reciprocity is the basic level of anthropological characteristics; and in the development of human information civilization, the importance of deepening our understanding of information characteristics at a specific mechanism level becomes evident. The created ex nihilo and sharability of information imply that the development of information civilization has an information resource foundation that is entirely different from that of matter and energy. The emergence of information signifies the creative driving nature of the development of information civilization, while reciprocity of information not only implies a deeper level of sociality but also the increasingly prominent development characteristics of human information civilization in relation to human characteristics of kind features.

Due to achieving a paradigm shift in information understanding, we can systematically derive the basic characteristics of information from the receptive relation understanding of information. The systematic derivation of the basic characteristics of information can not only conversely constitute mutual interpretation with the understanding of information but also allows for a more accurate understanding of human information civilization and its development.

8 Information and the Development of Information Civilization

8.1 Introduction

The receptive relation understanding of information and the systematic derivation of its basic characteristics provides a new theoretical foundation for deepening our understanding of human information civilization. The development of information technology has ushered humanity into a new form of civilization. The study of this new form of human civilization, from the "post-industrial society" to the "information society," from the "information age" to the "information civilization," has undergone a continuous process of deepening. Research on the "information age" and the "information society" has become increasingly thorough and systematic, and elevating the study to the level of information civilization is a recent trend. The development of information civilization unprecedentedly highlights two important aspects. On one hand, the study of information civilization cannot proceed without a deepened understanding of information. On the other hand, the deepening understanding of contemporary human social development cannot be without holistic guan zhao of information civilization.

Primarily based on industrial positioning, the "information society" is defined with reference to the "industrial society." "post-industrial society" reflects the transitional social form from industrial civilization to information civilization (Bell, 1973). People have analyzed information society from aspects such as information technology, occupational structure, economy, spatial structure, and culture, providing quantitative descriptions of the information society, "where more than half of the gross national product comes from the informational economy sector, and half of the employed population is engaged in activities related to the informational economy" (Bell, 1976). Although this discussion mainly focused on the paradigm of social production, especially economic development, it shifted the focus to the social positioning of the information society and the information age. The development of the internet has provided the most important foundation for transitioning from economic or industrial positioning to social positioning. Unlike industrial society, the fundamental basis of information society is the network.

In 1996, Manuel Castells introduced the concept of the "Information Technology Revolution" (Castells, 1999). This "ubiquitous in all human activities" revolution became his point of entry for "analyzing the forming new economic, social, and cultural complex systems." He argued, "As a revolution, the information technology

DOI: 10.4324/9781003484851-8

revolution emerged in the 1970s" (Castells, 2010). The information technology revolution opened the curtain on the Information Age. As early as the early 1980s, Alvin Toffler mentioned the concept of the "Information Age" (Toffler, 1991); the surge of the "Third Wave" precisely embodies the surging rhythm of the development of the "Information Society."

If the internet is the foundation of the information society, then big data is the foundation of the development of information civilization. With the development of big data, concepts of the "new information society" and the "new information age" have been proposed on the basis of the "information society" and the "information age."

On the new information society: "It is a world in which everything is measurable and in which people, and almost every device you can think of, are connected 24/7 through the Internet" (Klous & Wielaard, 2016, p. xiv). The emergence of the concepts of the "new information society" and the "new information age" has gone through a process marked by the appearance of new information technologies, initially centered on computers and later transitioning to a foundation based on big data. The new information society is built upon the foundation of big data, which is a fundamental component of this new information society. As big data continues to evolve, the concepts of the New Information Age and New Information Society increasingly transcend their information technology origins and further reflect information civilization as a brand-new form of human civilization.

Back in the 1990s, the renowned American scholar Mark Poster introduced the concept of the "mode of information" (Poster, 1990). Subsequently, several scholars conducted in-depth theoretical and practical research on information civilization. Farhang Rajaee of Carleton University in Canada proposed the concept of a globalized information civilization based on the development of information technology and the fusion of different civilizations (Rajaee, 2000, pp. 8–9). The research by Shoshana Zuboff in Harvard University, among others, not only criticized "surveillance capitalism" in defense of the spirit of information civilization but also addressed practical issues related to the pathways from the United States and Europe (Zuboff, 2014, 2015).

Starting from the philosophy of technology, Professor Feng Xiao conducted a systematic study of information civilization, asserting that the philosophy of technology, with its unique perspective, links the nature and appearance of society to the development and revolution of technology (Feng, 2015). Scholars like Sumei Cheng explored information civilization from the perspective of its connection to the economic and social realms (Sumei, 2018). Building upon the foundation of research on the "information society" and the "information age," the systematic in-depth exploration of both the philosophy of technology and socio-economics has laid the groundwork for the continuous deepening of research on information civilization. This has led to a shared understanding of information civilization as a form of human civilization that stands alongside agricultural civilization and industrial civilization.

Information civilization, alongside agricultural and industrial civilizations, is a categorization of human civilizations based on technological forms. Regarding this understanding of information civilization, Feng Xiao provided the most representative expression: "The fundamental perspective underlying the three major categorizations of civilizations, including 'information civilization,' is a technological perspective" (Feng, 2017). Based on the categorization of human civilizations according to technological forms, research on information civilization has established a basic framework. Building upon this framework, as the understanding of information and its relationship with the material world deepens; research on information civilization also continues to deepen.

By equating information with matter and energy, Wiener hinted at a unique shift from the paradigm of matter and energy to the paradigm of information, foreshadowing a new horizon in human development. Building upon Wiener's understanding of the relationship between information and matter/energy, it becomes evident that the nature of the relationship between information civilization and industrial or agricultural civilizations is fundamentally different from the relationship between industrial and agricultural civilizations. As human information civilization has developed, we have witnessed the unprecedented significance and manifestation of a new civilization, one unlike any that humanity has seen before. Information, which is neither matter nor energy but has dependencies on both, along with its fundamental characteristics, implies an entirely distinct form of human civilization.

In the sense that information is neither matter nor energy but still requires material attachments and energy for propagation, information civilization stands as a form of civilization distinct from largely material civilizations. The parallel status of information alongside matter and energy has led to a significant transformation in the way human societies are organized, with information activities taking an increasingly dominant role. This new mode of existence signifies a higher-level human civilization that has developed on the foundation of material civilizations.

This leads to a deeper understanding of information civilization: agricultural and industrial civilizations are primarily based on matter and energy, encompassing all previous stages of civilization development as phases of largely material-based civilizations. In contrast, information civilization is a new form of human civilization that stands parallel to the entirety of largely material civilization. This categorization is based on the foundational elements of human civilization, encompassing both matter and information (Tianen, 2015). This perspective can also be understood from the standpoint of atoms and bits: "The industrial age, very much an age of atoms" (Negroponte, 1995, p. 163) and this is even truer for agricultural civilization. In this understanding, agricultural and industrial civilizations are essentially "atomic civilizations," whereas information civilization is a "bit civilization." Through receptive relation understanding of information, it can be seen more clearly that the parallel between "atoms" and "bits" arises from an information encoding perspective. However, by surpassing the information encoding ceiling of information understanding, the theoretical logic becomes clear.

Undoubtedly, human civilization has always been an integrated civilization of matter and information. However, its structure is the development of information civilization based on material civilization. Therefore, the development of human civilization has become a process that shifts from primarily manifesting as material civilization to increasingly manifesting as information civilization. In the development of material civilization, information serves the development of matter and energy, where information is the means and matter and energy are the ends. In the development of information civilization, matter and energy serve the evolution of information, where matter and energy are the carriers, and the ends are the development of information and information agents. This Copernican reversal marks the turning point from material civilization to information civilization. The foundation of human civilization will always be matter and energy, but the essence of human civilization is information. In essence, its development process is the evolution of information, and the development of human society increasingly manifests as the development of information civilization.

8.2 The Informational Development of Human Civilization

The systematic understanding of information and its basic characteristics implies a deepening understanding of human information civilization: a higher-level information civilization largely based on matter and energy. Information civilization constitutes a higher-level facet of human civilization, providing a key to a deeper understanding of "development." The door opened by this key reveals the unfolding of human civilization through information.

8.2.1 From Material-Sharing to Informational-Sharing

Information civilization not only builds upon the shared natures but also greatly expands the shareability of matter and energy through the genuine shared nature of information.

On one hand, as information civilization develops, it becomes increasingly important to clarify the distinctions between material-sharing and informational-sharing. Unlike matter and energy, information possesses a natural propensity for sharing, it makes the difference between di-sharing and co-sharing. These two have entirely different implications for human civilization. With material resources, the more they are shared, the less each sharer possesses; however, in the case of information, the more it is shared, the more each participant gains. In the context of di-sharing, material civilizations often entail conflicts over material interests, whereas the inherent nature of informational sharing forms the foundation for co-sharing in human civilization.

The inherent sharing nature of information not only implies that it does not diminish due to sharing but may also stimulate more information as sharing expands. As information civilization advances, the inherent nature of the sharing of information will facilitate unprecedented possibilities for di-sharing material resources. The commonly observed trend of hardware becoming software-oriented serves to extend the potential for sharing matter and energy or space through information expansion. This is precisely why industrial civilization increases the possibilities for sharing matter and energy, while information civilization significantly extends the potential for sharing material resources.

However, as resources, matter/energy and information possess fundamental differences. Material resources can only be di-shared but lack the inherent co-sharing nature of information resources. Material resources not only do not increase with sharing but also deplete faster as sharing expands. This results in synchronic and diachronic sharing of combining matter/energy and information on the one hand, and diachronic di-sharing substantially on the other hand. In this sense, material resources are fundamentally di-shareable in nature. Although information civilization cannot change the diachronic di-sharing nature of material resources, it can increasingly unfold the synchronic di-sharability nature for combination of matter/energy information, providing a solid foundation for information civilization to fully develop and share material resources. More importantly, the development of information civilization will continue the informatization of matter and energy, thereby continuously expanding the synchronic sharing of material resources. DNA and microchips are respectively most typical examples of natural evolution and human creation. These involve important mechanisms of development from datafication to informatization of matter and energy.

As information civilization advances, the shared nature inherent in reciprocity becomes increasingly prominent. Reciprocity has evolved and expanded along with human development, representing a characteristic requirement for human sharing. Human nature fundamentally revolves around human needs, and human development essentially entails the development of these needs, each with its distinct nature. The lower-level satisfaction of needs tends to be individualistic, with material needs showcasing this typical characteristic. Conversely, as the level of needs rises, the unfolding of shared nature becomes more pronounced, making a higher level of material need satisfaction more communal. Lower-level needs bear the nature of matter and energy, while higher-level needs take on the nature of information. In the process of progressing from physiological needs to psychological needs and further to spiritual needs, matter and energy assume a more fundamental position at lower levels, while information occupies a position farther from the material foundation. The inherent differences between information and matter/energy determine the distinct characteristics of information civilization and material civilization.

Certainly, information cannot be separated from matter and energy, and the civilization of information must be based on material civilization. However, information civilization is not externally built upon the foundation of material civilization. Instead, it plays an increasingly important role in material civilization through the integration of civilizations. This transformation makes material resources increasingly the co-shared foundation of information civilization. The development of information civilization allows humans to control matter and energy through information, adjusting the structure of matter to make it more valuable and satisfying for human needs. It transforms matter and energy from something not easily obtained into something more accessible, enabling human activities to be more directly related to information rather than matter and energy. Due to the inherent nature of information sharing, human society has gradually entered a phase of co-shared development with the advancement of information civilization, even achieving widespread creative development. "Sharing opens the perspective of a many-to-all cultural society, in which everyone has access to creative, expressive and informative works and has the means to contribute to their creation" (Aigrain, 2012, p. 50). This not only relates to social progress but is also closely connected to human development. Therefore, the inherent nature of information co-sharing signifies the information unfolding of human civilization, leading to significant changes in ownership and usage relationships in co-shared civilization.

8.2.2 From Ownership to Usage

Information civilization has shifted the significance of resources from ownership to usage. It is the basic principle of sharing, where convenience and even greater convenience come from access without the necessity of ownership. This principle forms the existential foundation for the development of information civilization.

Issues related to access and ownership have long been discussed in fields such as library science, but the real debate regarding usage and ownership is relatively recent (Hoadley, 1993; Brown, 1995). This reflects a crucial trend in contemporary development. With the development of information civilization, more and more people tend to "access rather than ownership", therefore "access over ownership" (Dupuis & Noreau, 2016). The development of the sharing economy has led people to feel that "We have left the age of ownership and taken the first steps in the era of access, where companies are confronted with consumers who no longer want to own products" (Meyer & Shaheen, 2017, p. 105). This is closely related to population concentration; the denser the population, the more effectively the sharing mechanism works. "If population density raises the cost of maintenance, the more density increases, the more individuals will generally prefer to pay for access rather than ownership" (Schneider, 2017, p. 28). The shift from ownership to usage is rapidly changing people's lifestyles and ways of existence. "In the biggest cities in Europe, we see people giving up ownership of the car to switch to sheer usage" (Stephany, 2015, p. 133). This transition is even altering some fundamental concepts. In the United States, for some women, "ownership was a sacred part of the American Dream," but "this preference for access rather than ownership constitutes "a seismic shift in the American dream" (Stephany, 2015, pp. 28, 27). Faced with the "Reduced need for ownership," the concept of "disownership" has emerged (Stephany, 2015, pp. 11, 26), undoubtedly involving a transformation of civilization.

The development of information civilization not only emphasizes the more effective possession nature of using information resources but also changes people's relationships with the ownership and usage of material resources. Even for material resources, ownership often does not align with fundamental human needs compared to usage. If not the most efficient usage strategy, ownership of material resources can often become a burden. The co-shared information civilization highlights the redundancy of owning material resources: the smaller the range of resource owners, the higher the ownership cost, yet the efficiency decreases. When ownership has a limited scope concerning material resources, its value in usage broadens, leading to more significant resource wastage. For information, besides the legitimate rights obtained through innovation, asymmetric ownership might be a severe waste as it involves the development mechanism of the information civilization era.

Co-shared civilization possesses its own unique mechanisms of development, one of which is the contradiction between information symmetry and asymmetry. According to the "symmetry-theory of information," "information is grounded in asymmetry," "information is a way of abstractly expressing asymmetries," and "information is asymmetry" (Muller, 2007, pp. 2, 141, 143) and the mechanism through which information operates is the symmetrization of information and this fundamental relationship determines the basic contradiction between information symmetry and asymmetry. Information symmetry is a crucial measure of the shared development level of information civilization. Simultaneously, at the forefront of information production, information asymmetry is a vital mechanism ensuring information innovation and production drive. On one hand, the importance of information symmetry in resource allocation, even for material resources, is self-evident. Information symmetry is crucial for the market's resource allocation, which relies on information. On the other hand, a necessary and temporary information asymmetry at a certain stage is a crucial measure to protect innovation initiatives and continuously renew the indispensable driving force of social development. Without effective protection of innovative patents, what is gained is immediate existing benefits, while what is lost is the fundamental long-term drive. Therefore, as social information civilization advances, maintaining a reasonable balance between information symmetry and asymmetry will increasingly become a fundamental principle in unleashing more people's innovative drive, effectively promoting information production, and advancing the progress of information civilization. This principle not only concerns the social development of information civilization in society but also deeply involves the development of individuals in the era of information civilization.

The higher level human needs are, the more they possess the nature of information, thus making their reciprocity stronger. The stronger the reciprocity between needs and their satisfaction, the less significance individual or small-scale ownership of resources holds, and the more meaning is given to usage over ownership. This implies the holistic nature of higher-level human information needs, signifying a commonality in the satisfaction of these needs. Therefore, the higher-level information needs necessitate cooperation for mutual satisfaction. Either everyone gets the satisfaction they need, or no one does. This is the synchronous mutual agency in satisfying information needs. This synchronous mutual agency implies a further fact: not only is the satisfaction of information needs shared, but also the generation and development of these needs are common. Emotions and many social needs are already evidently mutual common needs among some people. While the formation of emotional needs and their relationships may involve a minority of individuals, the occurrence and development of intellectual production needs can involve an exponentially increasing number of people. The higher the level of the generation and satisfaction of information needs, the more likely they directly or indirectly involve more people, potentially encompassing an entire humankind, thus exhibiting a greater degree of commonality. This is because only in the synchronous mutual agency of humans can higher-level information needs arise.

As human information needs progress to higher levels, they become increasingly holistic within specific groups. The more humankind-like the nature of these needs, the more commonality their satisfaction holds. Due to this commonality in satisfaction, within specific groups, this holistic information need is not something certain individuals can satisfy independently while others are left unsatisfied. In this holistic information need, a new characteristic emerges: human information needs are not only mutual agency in the general sense but also each individual serves as an agent for the mutual need and its satisfaction. Only by representing and satisfying the holistic information need of humankind can individual information needs be satisfied. Therefore, the relationship between ownership and usage of resources will be entirely different, and as humans increasingly exist in an informational way, this change is greatly accelerated. Because of this, it's important to emphasize the fundamental characteristics of information civilization, further expanding contemporary "development" within the advancement of information civilization.

8.3 Basic Characteristics of Information Civilization

In line with the fundamental nature of information, information civilization possesses basic characteristics distinct from those of material civilization.

8.3.1 A Human Civilization Directly Based on Information

As a higher-level form of human civilization built upon the foundation of material civilization, information civilization must be comprehended at a higher theoretical level. On the one hand, this implies understanding information civilization within the intrinsic interconnectedness of information, big data, and artificial intelligence. On the other hand, it suggests that we must understand information, big data, and artificial intelligence within the holistic guan zhao of information civilization.

Big data possesses characteristics of scale integrity, structural openness, real-time flow and value production. It provides a quantitative means of comprehending the existing world within the context of information civilization, carrying profound ontological significance. The holistic guan zhao of information civilization with respect to artificial intelligence not only highlights the civilization implications of artificial intelligence but also unfolds an intelligent understanding of information civilization. The fundamental nature of information civilization is holistic informatization, and the high-level development of holistic informatization is holistic intellectualization. It is precisely as an important foundation for the development of holistic intellectualization in information civilization that artificial intelligence possesses profound information civilization implications.

Achieving a deeper understanding of information through the unfolding of information in big data and artificial intelligence allows us not only to understand information civilization based on the foundation of material civilization but also enables us to understand material civilization based on the foundation of information civilization. Moreover, it facilitates the understanding of the relationships between nature, society, and the world within the context of material civilization and information civilization. This deeper understanding allows us to reevaluate social progress and human development within the holistic guan zhao of information civilization. Additionally, within the holistic guan zhao of information civilization, it enables us to understand and grasp information civilization more comprehensively and effectively.

Due to the contemporary development of artificial intelligence, the information civilization is progressing toward a higher stage of its intelligent era. As the intelligent era is in the emerging stage, the current understanding of this concept and a precise definition of the "intelligent era," especially in terms of its nature and fundamental characteristics, have not yet reached a consensus. However, in the context of contemporary development where "the future is already here," building upon the existing foundation of progress, providing a preliminary framework for understanding the intelligent era undoubtedly holds important holistic guan zhao significance for both specific comprehension and societal practices.

The core driving force of the intelligent era is intelligence itself. The development of various aspects of human society, including the economy, politics, culture, and more, will predominantly rely on the holistic intelligence structure of society. This includes human intelligence, machine intelligence, networked composite intelligence, and so on. Simultaneously, profound changes will occur in the way humans produce, live, and interact. As the impact of material on humanity becomes increasingly foundational, the development of the information civilization is leading people to exist more and more in an informational way. Consequently, creative information work will take center stage, people will pay more attention to their information needs, and social interactions will expand to higher levels and encompass multiple dimensions.

8.3.2 A Co-Shared Civilization

When we classify civilizations logically, following the logic of agricultural civilization and industrial civilization, the information civilization will inevitably appear before us in a similar fashion. However, when we view past forms of civilization as primarily material civilizations based on Wiener's understanding of information, we may discover the following fact: material is increasingly taking on the nature of a carrier, and ultimately, the real protagonist is information. In this sense, this understanding of an information civilization is fundamentally more essential than the distinction of "information civilization" within agricultural and industrial civilizations. What does an information civilization, relative to a material civilization, imply? This question undoubtedly touches upon a more fundamental issue.

8.3.2.1 The Information Level of Human Civilization

Information is neither matter nor energy, so from the perspective of entities, information has the property of being attached to carriers. But if viewed from a practical perspective, considering that language, for instance, only has meaning in its usage, we encounter a reversed fact: due to its involvement in the forms of existence, information is fundamental even if we do not argue that the carrier itself is determined by information in some sense. In this sense, information holds a more important position than matter and energy. In the relationship between humans and the use of matter and energy, information is taking on an increasingly important role. It is in this sense that matter and energy have more importance than information, but information has more fundamental importance in use than matter and energy. That is to say, in the sense of usage, in the sense of big data being the fundamental platform of usage, the relationship between information and matter/energy indicates that information has more fundamental aspects than matter and energy. Because as carriers, material can be substituted, it is replaceable, and the same structure can be composed of one form of material or another, and the same function can come from a variety of different structures. The key factor here is not the specific material itself but the information. This understanding can be succinctly explained by genes in DNA: genes are more "information" than a "substance." The emergence of information means that, on different material foundations, products with the same function that satisfy the same human needs can be created. It is precisely this aspect that connects the characteristics of information with the information civilization itself. One of the most important connections is the inherent nature of information co-sharing.

The information civilization is a civilization of co-sharing. On one hand, information inherently possesses the quality of being co-shared. Unlike material resources, information only holds value when it is used, and its value is realized almost entirely through co-sharing. The broader the scope of information co-sharing, the greater the value derived from it. Consequently, the unit cost of information decreases as the scope of sharing expands. In other words, information sharing tends toward zero marginal cost because information resources are used in coded form, with digital coding being the most crucial method, measured in bits. Unlike atoms, bits are easily duplicated and propagate quickly without time or space constraints. On the other hand, the development of the internet, especially artificial intelligence, provides the technical means for information sharing. Information technology, as a tool for processing and disseminating information, inherently needs information sharing. Without sharing, information cannot be processed, transmitted, or realize its value. The internet serves as a full manifestation of this technical logic for information sharing. The core value of the internet lies in equal interaction, diverse freedom, and open sharing of information. With the advent of mobile internet, more and more people are connected to the network, enabling them to share information anytime and anywhere. Information sharing is increasingly becoming a part of everyday life. Thus, the inherent nature of information sharing, coupled with the technical means provided by the internet, has facilitated the arrival of the information civilization era, which fundamentally is a civilization of co-sharing. Furthermore, as human information civilization evolves, the development of the Internet of Things (IoT) connects more and more material resources to the network, greatly expanding the possibilities of di-sharing matter and energy through information.

Nowadays, shared bicycles, shared power banks, shared umbrellas, and similar services are practically ubiquitous and can be accessed with standard information devices. In a smart environment, the space for sharing can expand exponentially. Although the sharing of material resources at its essence still is a kind of diachronic sharing that ultimately involves a time-bound, the inherent nature of information sharing sets it apart from previous material civilizations. The co-sharing aspect of the intelligent era progresses to a higher stage – from the sharing of information encoding to the strengthening of reciprocity of intelligent agents, exemplified by mutual stimulation.

(2) The Basis of Information in a Co-Sharing Civilization

As a result of the economic laws inherent in information sharing, the decreasing marginal cost of replicating and disseminating information products establishes the economic basis for information sharing. The network effect, as a technological law fostered by mobile internet, lays the technical foundation for information sharing. It is these laws based on the nature of information that give rise to a co-sharing civilization distinct from past material civilizations.

Because of the inherent nature of information sharing, it fundamentally alters the situation of material interest conflicts that have always accompanied (if not intensified) material civilizations. This creates a basis for the expansion of co-sharing on a principle that can theoretically extend infinitely in human civilization development. Due to the economic laws shaped by the nature of information, information products become increasingly cost-effective to replicate and distribute, making information sharing not only conflict-free but also potentially infinite in scope. Information civilization is a type of civilization where, as the scope of sharing expands, unit costs continuously decrease. Therefore, meeting the needs not only won't become a burden – （just like satisfying increasingly more people's eating needs will be a disastrous ending), but instead, it becomes a mutual relief, forming a more robust mutual support mechanism – much like how more users reduce the cost of software production. In contrast, the sharing of material resources becomes increasingly burdensome as the number of sharers increases, whereas information sharing, on the contrary, makes sharing costs decrease as the number of sharers grows.

(3) The Humanity Basis and Philosophical Foundation of Co-Sharing Civilization

Co-sharing civilization not only requires material conditions created through information but also has its basis in human development, thus having its philosophical foundation. In informational civilizations every subject becomes agent, therefore, in increasingly more use situations the tendency for "agent" to replace "subject" is an indispensable concept and also is the necessary situation condition of further clarifying this concept. In the information civilization era, due to developing to a certain level, machine intelligent agents and even programs all are agents, therefore when dealing with the relationship between human and artificial intelligence this concept not only is indispensable but also increasingly important.

"Agent" is a highly contextual concept, and understanding it becomes a significant philosophical question. Understanding of agent most closely related to causation should be "an active and efficient cause." In philosophy, an agent generally refers to an entity which is capable of action, has agency, and is an intermediary, factor, actor, doer, and motive force and so on.

The more reasonable understanding of the "agent" concept must ascend to the information level. When dealing with physical entities, information is one of the interacting factors, but it is relatively dynamic and exerts a more dominant influence. This is an extreme way in which information acts. Another extreme way in which information acts is in human interactions. In human interactions, there is not only interagentivity (from intersubjectivity) but also a mirrored relationship. Interagentivity is typically manifested as mutual agency, meaning that the relationship between information agents is reciprocal agency. This mirrored relationship implies that information agents not only mutually depend on each other in terms of interaction but also act as agents for each other in terms of effects and benefits.

In essence, information agents mutually satisfy each other's needs, especially when these needs develop to higher levels. In basic, free-of-charge services and value-added services charged, commonly called "wool comes out on pig," through the cross subsidization of attention clustering, this reciprocal agency is a primary form of market behavior.

In lower-level activities, even though the satisfaction of needs may be conflicting, mutual agency is sometimes necessary to meet the needs of each other. In higher-level activities, the needs of any information agent must be met through cooperation to achieve common satisfaction. This satisfaction is no longer individual but common. Either all parties are satisfied, receiving the needed satisfaction, or none are satisfied. This is a typical mutual agency in need satisfaction.

Furthermore, not only is the satisfaction of needs common, but the germination of information needs is also common. In emotional and social needs, for example, people essentially have mutual common needs. Only in the mutual agency of information agents can higher-level needs arise. As human needs develop to higher levels, they become more humankind-based rather than individual needs that some people have and others do not or do not need. In these humankind-based needs, a new characteristic emerges: the needs of information agents are not only mutually agentic but each information agent is an agency for the common needs and their satisfaction. Only by representing and satisfying humankind-based needs, which are the needs of everyone, can true self-satisfaction be achieved.

8.4 An Information Civilization Unfolding of Contemporary Development and Its Foundations

The advancement of human civilization through information not only brings forth new ideas for the development in the era of information civilization but also involves further unfolding of human existence and even the dynamics of the human ecological system and societal progress.

8.4.1 Unfolding of Human Existence and Development

Information civilization further unfolds the ways in which humans exist. Based on big data, the fit between informatization of material and the information way of human existence is another basic fact of human information civilization. This not only establishes a higher-level foundation for existence and development in the era of information civilization but fundamentally alters how humans exist. As information civilization progresses, humans will increasingly exist in an informational way. The cyclical process of informatization of material and materialization of information implies not only the informatization of existence related to humans but also the informatization of human existence itself. This further expands the fundamental characteristics of anthropology within the context of information civilization.

The development of information civilization further underscores the fundamental nature of humans as a kind of social relationship, with social development being manifested as the genuine unfolding of reciprocity, thus leading to an unprecedented growth in human information needs. The development of information needs, in turn, signifies that human development enters into a virtuous cycle of informatization.

As beings in an informational way, humans have increasingly crucial information needs. "'Information need' is becoming the 'primary need' for more and more people in terms of both time and energy" (Feng, 2017). Moreover, the more humans exist in an informational way, the stronger their need for information becomes compared to their need for material resources. Due to the growing association with the informational way of human existence, information technology will continue to have a profound impact on the development of human information needs. Recent studies even term the era of information civilization as the "knowledge civilization era," contending that "the technology of knowledge civilization era will differ from that of industrial era in proposing boundless number of diversified technological possibilities" (Wierzbicki, 2006). Information technology further highlights the dual nature of technology as a double-edged sword, bringing about both benefits and new challenges for humanity. Studies suggest that the drawbacks of the new information society are becoming increasingly evident, including privacy breaches, lack of transparency, and making money from personal data (Klous & Wielaard, 2016, pp. 19–22). Dealing with the pros and cons of information technology development is contingent upon a higher-level understanding and response within the framework of human civilization.

As information civilization evolves, individuals and society will undergo a series of changes, fundamentally reflecting the development of human needs. A general trend is that material needs will gradually diminish, some of them even shrink and disappear, while informational needs will correspondingly undergo rapid development. In this process, there will be a significant shift in the emergence of new needs, with new material needs virtually non-existent, and new informational needs arising more and more. This change signifies both development and the corresponding regression.

The shift in human needs and their development will give rise to a series of transformations in values. Therefore, the capacity for reflection and critical thinking in individuals and society is becoming increasingly crucial. This involves both the nature of human informational needs and the determination of information value based on informational needs.

The state of existing as informational beings depends on the satisfaction of information needs. One surprising aspect for individuals existing in an informational way is that the satisfaction of information needs has a more pronounced impact than the satisfaction of material needs does for individuals primarily existing in a material way. Viewing this from the opposite perspective might make it clearer. The effects of temporary and minor discomfort from a surplus of food on the human body are relatively inconsequential, whereas an imbalance in the information system may have a fatal impact for people who exist mainly in an informational way.

As people who mainly exist in an informational way, an internal imbalance in the information system fundamentally relates to the state of information production, involving the formation and performance of creativity. In the era of information civilization, the interplay between the performance of human creativity and human existence in an informational way constitutes an unprecedentedly prominent bidirectional cyclical mechanism. It is precisely this bidirectional constructive relationship that further expands the informational layer level of human ecology.

8.4.2 The Information Civilization Unfolding of the Driving Mechanism of Human Social Development

The development of the information civilization has further unfolded the driving mechanisms of human social development. Due to the importance of matter and energy in quantity, basic physical activities are mostly repetitive production activities. On the other hand, due to the qualitative significance of information, basic information activities lean more toward creative production activities. Therefore, the development of the information civilization has made innovation increasingly critical to progress. In the age of information civilization, creativity is becoming the core driving force for development. As the driving force behind the development in the information age, innovation depends on human creativity. On an individual level, a person's creativity is influenced by their intelligence, emotional intelligence, and practical intelligence. On a societal scale, a society's capacity for innovation relies on a country's education system and overall cultural level. From the perspective of the individual-society relationship and the actual outcomes, a society's creativity depends on the extent to which individual creativity is liberated in that specific society. Innovative development is tied to the performance of human creativity, which depends on the widespread liberation of people's creativity. This widespread liberation, in turn, is linked to the development of individuals' higher-level needs.

The drive for innovation itself is intrinsically linked to the development of human needs. Purely technological innovation is more easily driven by material needs, as material interests can be a significant driver of technological innovation since it's more directly related to material gains. However, the more original thought production is, the more indirect its relationship with material interests; the more basic thought originality is, the harder it is to be driven by external material interests. Basic theoretical and ideational innovations, which are often very foundational in nature, are challenging to achieve through external material incentives and instead require inner motivation originating from higher-level needs. From real societal development, it can also be seen that the realizations of larger-scale technical engineering innovation not only can but also need to be achieved through the driving force from societal organizations. This kind of development can be relatively detached from the development of individuals themselves and may even run counter partly to the development of individuals. However, the more original the development of fundamental theories and innovative ideas, the more they depend on inner motivation, which, in turn, hinges on the widespread development of individual needs. Moreover, the drive associated with higher-level inner needs greatly depends on the overall cultural and civilizational level of society. Without the development of individual needs, individual development, and an elevation in the overall cultural and civilizational level of society that aligns with individual development, widespread achievement of fundamental theoretical and ideational innovations is unlikely.

A country that lacks sufficient foundational theories and original thoughts to support technological innovation cannot sustain such innovation, except by riding on other nations' thought production public bus. However, relying on another country's intellectual production is unsustainable for sustaining technological innovation. This is because original foundational theories and innovative ideas are not just the foundation for technological innovation but, as one of the most crucial foundational conditions, they determine the overall cultural level of a society. Consequently, this influences the level at which a society can liberate and perform its creativity. This involves a more complex bidirectional relationship. Not only is overall societal culture the foundation of thought production, but thought production itself is becoming an increasingly critical foundation for human civilization. While this aspect might have been somewhat obscured during the material civilization era, it is now more pronounced in the era of information civilization. In which, one of the most crucial benchmarks of overall societal culture is the informational ecosystem. The information civilization unfolding of the developmental mechanism suggests that the overall societal culture and thought production mutually support each other in this era. These two components constitute a bidirectional cyclical relationship, which is the core developmental mechanism of the information civilization.

Human development is a bidirectional cyclical process involving individual and societal development. Throughout the course of human development, the bidirectional cycle comprising individuals and society is based on matter and energy, with information taking on a higher level. In the initial stages of human social development, "the importance of information resources is far less than that of material resources" (Feng, 2017), and the physical cycle predominates. However, as human society advances further, information's pivotal role becomes increasingly pronounced as material resources take a more foundational position. The process of human social development is marked by information occupying a higher-level interface position, signifying that individual development, as the fundamental information agent, is increasingly prioritized.

In the bidirectional process of personal and societal development, the relationship between individuals and society undergoes a covariation due to different stages of development. In the era primarily based on material resources, known as the material civilization, societal development took precedence over individual development in this bidirectional process because of the dominant position of material resources. Without societal development, there couldn't be a foundation for individual development to take place. However, in the era primarily based on information, known as the information civilization, individual development takes precedence due to the intellectual advantage of information resources. Without widespread individual development, there can be no normal societal development, let alone rapid economic growth. It's within the framework of individual development in this context that we can not only deepen our understanding of liberalism and communitarianism and their relationships but also recognize a fundamental aspect highlighted by the information civilization: the nature of "development" itself.

8.5 Information Civilization Highlighting Human Development

The nature of the information civilization suggests that over a considerable period of history, the development of human society has shifted from primarily existing directly based on matter and energy to primarily existing directly based on information. The mechanisms of development are intrinsically related to the evolving needs of individuals. In this sense, the information civilization emphasizes the fundamental point: development ultimately boils down to human development.

8.5.1 The Basic Mode of Social Development is from Quantitative Increase to Qualitative Progress

Development can be about quantity increase or quality advancement. Quantity increase is a prerequisite for quality advancement, but it doesn't necessarily lead to it. In contemporary societal development, mere GDP growth can be a form of quantity-based development devoid of progress. True progress, on the other hand, involves not only economic development but also intellectual production, reflection on actions and thoughts, and the elevation of the levels of human and societal needs. Real development always includes progress, and progress signifies an increase in the levels of human and societal needs. Agricultural civilizations often remained in a state of quantitative development without qualitative progress because the needs of peasants not only always stay in their one mile lands but also are satisfied with the cycle of making money by herding sheep, then getting married and having children, and then herding sheep when the children grow up, generation after generation. This was not just a one-way closed loop but also a lack of fundamental reflection. The development of industrial civilization differed from agricultural civilization due to the increase in information factors, and the development of the information civilization differs fundamentally from the entire material civilization.

Development in the information civilization era is centered around progress, and without progress, there is no genuine development. The inherent nature of information co-sharing means that information products, in particular, become more competitive the more their content is information-based. For products primarily based on material resources, such as food, repetition and accumulation of quantity are crucial. In contrast, for products primarily informational, like computer software, quality renewal and upgrading are paramount. Replicating information products is not the issue; the challenge lies in renewing and upgrading them. This means that there is a significant risk associated with information product renewal, which is entirely different from material products. Before reaching a certain level of production and cost efficiency, a more advanced material tool wouldn't immediately render a relatively outdated tool obsolete. However, the appearance of a higher-level information product could quickly make lower-level relative products completely obsolete. Electric saws existed alongside hand saws for a long time, but once smartphones appeared, devices like pagers and Walkmans could only find their place in museums.

If we say that the development of material products is more related to meeting people's preexisting needs, then the updating and upgrading of information products are based more on the development of human needs. This means that it is more connected to the reflection on old products and related concepts, indicating that social progress is closely related to the reflection of the prerequisite foundation. The higher the stage of development, the more certain it is that the level of human and societal needs will rise, which is a result of reflection of the prerequisite foundation on established concepts, ways of thinking, and acting in practice. The most important, or rather, genuine reflection is a prerequisite reflection of concepts, theories, and ways of thinking, acting, and even ways of existing. This is also the significant reason why the development in the information civilization era signifies the integration of philosophy, technology, economy, society, culture, and even life. Progress in development comes from prerequisite reflection; progress always takes place and is achieved through prerequisite reflection, resulting in the updating of fundamental concepts, whether in science or social development. Einstein did not overthrow Newtonian mechanics; instead, he extracted more or higher-level specific conditions with new fundamental concepts, allowing the development of physics to exceed the scope of Newtonian mechanics. What he overturned were the fundamental concepts or assumptions on which Newton established classical physics. In the social realm, better living conditions will prompt the pursuit of higher-level meanings of life, leading people's interests to increasingly turn toward more innovative work, and this is the driving force behind innovation-driven development.

Economic development, driven fundamentally by innovation, is undoubtedly interconnected with the internal realm of thought production. The more human civilization is primarily based on material, the more societal economic development is a prerequisite for thought production. Without the corresponding material production base, thought production is not possible; however, in the human civilization era primarily based on information, the situation gradually reverses. In the information civilization era, development is premised on the progress of concepts and thoughts. Without the development of concepts and thoughts, the development of the society and economy will be greatly limited. With the development of the information civilization, the development of the society and economy and the development of thought production constitute a bidirectional cycle. In this bidirectional cycle, thought production plays an increasingly important dynamic role, closely linked to the process of human development.

8.5.2 *Human Development: A Process from External Conditions to Inner Needs*

The increasingly apparent trend of integrated development in the information civilization era is not just a phenomenon; it relates to an important aspect of the internal mechanism of development. This mechanism requires incorporating development into the overall consideration of human development. Since human development is the development of human needs, the development in the information civilization era must comprehensively promote the development of human needs. The development of moral needs is the most typical, as it is not only an important aspect of human need development but also a social reality that involves overcoming the moral decline resulting from economic development.

Examining the fundamental approach to the development of social morality from the perspective of the information civilization can better understand the inherent root of the moral decline in economic development. An important reason for the moral decline in economic development is closely related to the development of human needs and its guidance. Adjusting the holistic social structure timely with the growth of the economy to adapt and promote the elevation of human inner needs is a crucial guarantee for the harmonious development of economy and morality. From the perspective of human creative labor, we can have a clearer view of the intrinsic relationship between economic and social progress and human development.

As the process of labor is also the process of the worker's development, the development of the worker implies an increase in the level of their labor. The more fully a person unfolds, the higher the level of their activities, indicating a higher level of labor and a higher quality of labor. This reveals an important upward spiral bidirectional cyclical mechanism: labor is the original driving force of social development, and at the same time, it is the unfolding of human beings; the more comprehensive a person unfolds, the higher the level of their activities, which also means a higher level of labor and higher labor quality. Consequently, we can see a crucial bidirectional interactive relationship between labor and human development.

On the one hand, the quality of labor improves with the development of the individual, and labor is the original driving force of social development. The sustained development of society fundamentally relies on the continuous improvement of labor quality. On the other hand, the development of society ultimately stems from the development of human beings, although the development of human beings must be realized in a social context. The deep-rooted basis for society to prioritize humans is that development ultimately stems from human development. The development of individuals and society forms a bidirectional cycle, where the absence of human development leads to the absence of societal development, and conversely, the absence of societal development results in the absence of human development. In different stages of societal development, the two directions of the cycle of human and societal development hold different positions.

In the Information civilization age, creative labor, as the most in harmony with human nature, is increasingly becoming a primary need for people. This not only implies the sharing of the results of creative labor but also relates to the higher-level unfolding of individuals. Here, a crucial, principle-based turning point is the shift from external conditions to human inner needs in the development process. What is becoming increasingly evident is that development fundamentally boils down to human development. Regarding human needs and their development, this should also be considered a fundamental criterion. The emergence of "makers" in the information civilization age is a noteworthy phenomenon that exemplifies the development of creative labor as a hallmark of human development (Hatch, 2013, p. 16). The labor of makers is not only a creative activity but increasingly reflects the full unfolding and free development of individuals, and it is a type of labor characterized by its sharing nature. This labor is the driving force of information civilization age development, and for a country reaching a medium income level, creative labor that reflects human development is indispensable for further progress.

8.5.3 A Bidirectional Cycle of Forming Upgraded Driving Force and Human Needs Development

The way people exist in the information civilization age further highlights the increasingly direct connection between the driving force of development and the inner development of human needs. The more closely it is related to human inner needs and their development, the stronger the driving force for societal development; conversely, the farther laborers' work is from what people are most concerned about, the less driving force societal development has. In the conditions of material civilization, the connection between the driving force of development and the development of human needs may not only be indirect but also potentially conflicting. In the information civilization age, characterized primarily by innovation information activities, the connection between the driving force of development and the development of human needs is becoming increasingly direct and aligned. Therefore, it is crucial to understand and continually promote societal development within the context of ongoing human needs development.

Human information civilization implies that the well-being of people is increasingly linked not only to the satisfaction of material life needs but also to the satisfaction of higher-level needs with economic and societal development. Whether a person feels happy primarily depends on the state of their most immediate needs or the possibility of satisfying them. Under normal circumstances, a person's most immediate needs are their highest-level needs. In the realm of societal development, economic progress enables people to gradually move away from simple physical labor, and labor for mere survival no longer provides the complete meaning of life.

The development of human information civilization, on the one hand, will undoubtedly cause people to lose the main sources of happiness based on material civilization. On the other hand, it will also give rise to new sources of happiness based primarily on information civilization. This change will manifest very intricately in people's actual lives. Though it overall reflects the development of human needs, particularly the evolution of intelligence, it may also lead to negative consequences in some aspects.

As far as the development of information civilization impact on humans is concerned, the richness and vividness based on the material nature are fading or disappearing. Essentially, the question of whether the information civilization age will make us happier involves the elevation of human well-being. Human development has moved from natural evolution to autonomous evolution, and the evolutionary process involves giving up and gaining. What to give up and what to gain must rely on a long-term perspective, and whether the perspective is long-term fundamentally involves the level of comprehensive insight. Therefore, as a provider of the highest-level comprehensive insight function, the significance of philosophy will become increasingly pronounced with the development of the information civilization age.

In the material civilization era, the situation of people serving the material was overshadowing human development, while in the information civilization age, where the material is enslaved by people, the prospects for human development are opening up. The phenomenon of makers is by no means a historical accident or the result of individual interests and hobbies; the maker movement is closely related to the elevation of human needs and is a natural outcome of historical development under normal social conditions. The emergence of the "new generation of innovators" reflects the developmental nature that is becoming increasingly prominent in the information civilization age. Humans are an existence as a special kind, and the higher the level of development of human needs, the more informational these needs are, and the more social they become. The content and level of societal development will also continue to rise as human needs develop. It is in this process that societal development and human development form a virtuous cycle at higher levels.

8.6 Conclusions

From the "post-industrial society" to the "new information society," from the "information age" to the information civilization, a new kind of human civilization is gradually unfolding. The receptive relation understanding of information and the basic characteristics of information systematically derived from this system show that human civilization is fundamentally different when it is directly based on information as opposed to being directly based on material. As a kind of human civilization directly based on information, information civilization has surpassed the development logic of agricultural and industrial civilizations, becoming a higher-level human civilization alongside various concrete forms of material civilizations that came before it. As the information unfolding of human civilization, its development progresses from matter and energy to information, from di-sharing to co-sharing, from ownership to usage, from materialization to informatization, and then to intellectualization, humans are shifting from a state of being enslaved by material, to a state of enslaving material by using information through the development of artificial intelligence.

The basic characteristics of information, different from matter and energy, determine that information civilization or the information level of human civilization implies a co-shared civilization, a civilization of creating something from nothing through the creative construction of information, a civilization that is more in line with human nature, and a civilization in which the meaning of human life increasingly derives from activities related to information creation.

In the development of the information civilization era, as humans increasingly exist in an informational way, and due to the difference in the nature of human information needs compared to material needs, human development is returning to a stage where individual development plays a central role in driving social progress. In the unfolding of information civilization, the mechanism of individual development for social progress has shifted from being primarily driven by external needs related to material to being increasingly driven by inner needs related to information. The difference in the levels of human information needs and material needs has become the fundamental driving force behind the development of human society at a deeper level.

Information civilization highlights human development like never before, from quantitative growth to qualitative progress. The basic mode of social development reflects the progression of human development from external conditions to inner needs, thereby forming a bidirectional cycle of development and the development of human needs. As a result, human development is ultimately fully expressed in the era of information civilization.

Social development is a form of human existence and evolves social development is the group form of unfolding way of human existence and development, and the dimension of informational ecology, based on the dimensions of natural ecology and social ecology, constitutes the space in which humans increasingly exist in an informational way.

From the holistic guan zhao of information civilization, due to the inherent nature of information co-sharing and the fact that information civilization is a creative construction of information civilization, contemporary development is witnessing a bidirectional cyclical acceleration of human development and social progress. In this bidirectional cycle, as material civilization reaches a certain level of development, and as information civilization is more in line with human nature, human development increasingly takes a priority position, and there is a more direct mechanistic relationship between the renewal of driving of development and the development of human needs. Because humans increasingly exist in an informational way, and because information civilization presents a superimposition of information reciprocity and anthropological reciprocity, the holistic level of social progress is becoming increasingly crucial for development. The level of informational ecology in a society and the degree of liberation of people's creativity are increasingly becoming determining factors in contemporary development. These conclusions drawn from the holistic guan zhao of information civilization hold significant importance for contemporary development of human society.

From information to information civilization, the paradigm shift in information offers a completely new space for addressing existing problems, raises new questions, and even opening up important new areas of research. The systematic deepening of information, as receptive relation, research involves not only the development of information technology but also the internal mechanisms of human civilization progress. Within this, one can see various complex bidirectional cycles based on information feedback. From the core mechanisms of general artificial intelligence to the mysteries of consciousness and life, from the information understandings of important philosophical orientating thought such as phenomenology, and hermeneutics as well as philosophy of language to the information theory of knowledge that integrates them, all are interconnected within these complex bidirectional cyclical mechanisms. For information as receptive relation, these are all crucial subjects that urgently require further systematic and in-depth research.

8.7 Acknowledgment

We are deeply grateful to *Social Sciences in China* for publishing the paper "A New Understanding of the Information Civilization 'Key' to 'Development'" (No.1, 2022) and *South China Quarterly* for publishing the paper "On Information Civilization" (No. 3, 2005).

References

Aigrain, P. (2012). *Sharing: Culture and the Economy in the Internet Age*. Amsterdam: Amsterdam University Press.

Avery, J. S. (2022). *Information Theory and Evolution*. Third Edition, Singapore: World Scientific Publishing Co. Pte. Ltd.

Barbieri, M. (2016). What Is Information? *Philosophical Transactions of the Royal Society. Series A, Mathematical, Physical, and Engineering Sciences*, 374(2063), 1–10.

Barwise, J. (1986). Information and Circumstance. *Notre Dame Journal of Formal Logic*, 27(3), 324–338.

Barwise, J. and Seligman, J. (1997). *Information Flow: The Logic of Distributed Systems*. Cambridge: Cambridge University Press.

Bawden, D. and Robinson, L. (2013). "Deep Down Things": In What Ways Is Information Physical, and Why Does it Matter for Information Science? *Information Research*, 18(3), 19–22.

Bei, Z., Xie, C., Duan-Lu, Z. and Xiao-Gang, W. (2019). *Quantum Information Meets Quantum Matter: From Quantum Entanglement to Topological Phases of Many-Body Systems*. New York: Springer.

Bell, D. (1973). *The Coming of Post-Industrial Society: A Venture in Social Forecasting*. New York: Basic Books.

———. (1976). The Coming of the Post-Industrial Society, *The Educational Forum*, 40(4), 574–579.

Béranger, J. (2018). *The Algorithmic Code of Ethics, Ethics at the Bedside of the Digital Revolution*. London: ISTE Ltd and John Wiley & Sons, Inc.

Beynon-Davies, P. (2011). *Significance: Exploring the Nature of Information, Systems and Technology*. Palgrave: Macmillan.

Bohr, N. (1932). *Atomic Theory and the Description of Nature*. Woodbridge, CT: Ox Bow Press.

———. (1958). *Atomic Physics and Human Knowledge*. New York: John Wiley & Sons.

———. (1963). *Essays 1958–1962, On Atomic Physics and Human Knowledge*. New York: Richard Clay and Company Ltd.

Boisot, M. H. and Canals, A. (2007). Data, Information, and Knowledge: Have We Got It Right? In: M. H. Boisot, I. C. MacMillan and K. S. Han eds., *Explorations in Information Space: Knowledge, Agents, and Organization*. Oxford: Oxford University Press Inc.

Bridgman, P. W. (1958). *The Logic of Modern Physics*. New York: Macmillan.

Brown, R. (1995). Access vs Ownership: Access Where? Own What? -- A Corporate View. *Serials*, 8(2), 125–129.

Bub, J. (2005). Quantum Theory Is About Quantum Information. *Foundations of Physics*, 35(4), 541–560.

Burgin, M. (2009). *Theory of Information: Fundamentality, Diversity and Unification*. Singapore: World Scientific Series in Information Studies.

Castells, M. (1999). *Flows, Networks, and Identities: A Critical Theory of the Informational Society, Critical Education in the New Information.* Lanham, Maryland: Rowman & Littlefield Publishers, Inc.

———. (2010). *The Rise of the Network Society.* Pondicherry, India: John Wiley & Sons Ltd.

Chalmers, D. J. (2010). *The Character of Consciousness.* Oxford: Oxford University Press.

Changlin, L. 刘长林. (1985). Lun Xinxi De Znexue Benxing 论信息的哲学本性. *Social Sciences in China* 中国社会科学,6 (2), 103–118.

Charles, P. (2000). *Code: The Hidden Language of Computer Hardware and Software.* Redmond and Washington: Microsoft Press.

D'Ariano, G. M, Chiribella, G. and Perinotti, P. (2017). *Quantum Theory from First Principles: An Informational Approach.* Cambridge: Cambridge University Press.

Dan, Z. (2019). *Phenomenology: The Basics.* London and New York: Routledge.

David, B. and Lyn, R. (2022). *Introduction to Information Science.* Second edition. London: Facet Publishing.

Dretske, F. I. (1982). *Knowledge and the Flow of Information.* Cambridge: The MIT Press.

———. (1983). Précis of Knowledge and the Flow of Information. *The Behavioral and Brain Sciences*, 6(1), 55–90.

Dupuis F. and Noreau, J. (2016). The Sharing Economy: A Black Box. *Desjardins Economic Studies*, 26(Oct.), 1–4.

Duwell, A. (2010). Schumacher Information and the Philosophy of Physics. In: A. Bokulich and G. Jaeger eds., *Philosophy of Quantum Information and Entanglement.* Cambridge: Cambridge University Press.

Eigen, M. (1971). Selforganization of Matter and the Evolution of Biological Macromolecules. *Naturwissenschaflen (The Science of Nature)*, 58(10), 465–523.

Feng, X. 肖峰. (2015). Jiyu Jishu Zhexue Shiye De Xinxiwenmin Tezheng 基于技术哲学视野的信息文明特征. *Journal of Northeast University (Social Science Edition)* 东北大学学报（社会科学版）, 17(1), 1–7.

———. (2017). Xinxiwenmin De BentilunJiangon 信息文明的本体论建构. *Philosophical Analysis*哲学分析, 8(4), 4–17.

———. (2018). *Xinxi De Zhexue Yanjiu* 信息的哲学研究. Beijing: China Social Sciences Publishing House.

Floridi, L. (2004). Open Problems in the Philosophy of Information. *Metaphilosophy*, 35(4), 554–582.

———. (2011). *The Philosophy of Information.* Oxford: Oxford University Press Inc.

———. (2019). *The Logic of Information: A Theory of Philosophy as Conceptual Design.* Oxford: Oxford University Press.

Fresco, N. and Wolf, M. J. (2016). Information Processing and Instructional Information. In: L. Floridi ed., *The Routledge Handbook of Philosophy of Information.* London and New York: Routledge.

Gleick, J. (2011). *The Information: A History, a Theory, A Flood.* New York: Knopf Doubleday Publishing Group.

Hardy, L. (2005). Why Is Nature Described by Quantum Theory? In: J. D. Barrow, P. C. W. Davies and Jr C. L. Harper eds., *Science and Ultimate Reality: Quantum Theory, Cosmology, and Complexity.* Cambridge: Cambridge University Press.

Hartey, R. V. L. (1928). Transmission of Information. *Bell System Technical Journal*, 7(3), 535–546.

Hatch, M. (2013). *The Maker Movement Manifesto.* New York: McGraw-Hill.

Hawking, S. (2005). Information Loss in Black Holes. *Physical Review D: Particles and Fields*, 72(8), 344–347.

Heisenberg, W. (1979). *Philosophical Problems of Quantum Physics.* Translated by F. C.Hayes. Woodbridge, CT: Ox Bow Press.

Hoadley, I. B. (1993). Access vs. Ownership: Myth or Reality – ScienceDirect. *Library Acquisitions: Practice & Theory*, 17(2), 191–195.

Hofkirchner, W. *The Logic of the Third: A Paradigm Shift to a Shared Future for Humanity*, Singapore: World Scientific Publishing Co.

Janich, P. (2018). *What Is Information?*. Minneapolis: University of Minnesota Press.

Jozsa, R. (2004). Illustrating the Concept of Quantum Information. *IBM Journal of Research and Development*, 48(1), 78–85.

Kelly, K. (1995). *Out of Control: The New Biology of Machines, Social Systems and the Economic World*. New York: Basic Books.

Khosrow-Pour, M. (2007). *Dictionary of Information Science and Technology*. Hershey PA: Idea Group Reference.

Khrennikov, A. (2004). *Information Dynamics in Cognitive, Psychological, Social and Anomalous Phenomena*. Dordrecht: Springer-Science+Business Media, B.V.

Klous, S. and Wielaard, N. (2016). *We Are Big Data: The Future of the Information Society*. Amstelveen: Atlantis Press.

Kun, W. 邬焜. (2017). *Bianzhengweiwuzhuyi Xin Xingtai* 辩证唯物主义新形态. Beijing: Science Press.

Landauer, R. (1991). Information Is Physical. *Physics Today*, 44(5), 23–29.

———. (1996). The Physical Nature of Information. *Physics Letters A*, 217, 188–193.

———. (1999). Information is a Physical Entity. *Physica A: Statistical Mechanics and its Applications*, 263(1–4), 63–67.

Laura, F. (2018). Quantum Theory Is Not Only about Information. *Studies in History & Philosophy of Science*, (72), 256–265.

Lombardi, O. (2004). What Is Information. *Foundations of Science*, 9, 105–134.

Lombardi, O., etc. eds. (2017). *What is Quantum Information?* Cambridge: Cambridge University Press.

Longo, G. (1975). *Information Theory: New Trends and Open Problems*. New York: Springer-Verlag Wieni.

Lynda, D. and Paul, L. (1991). *Information in Action: Soft Systems Methodology*. London: Macmillan Education LTD.

Mattingly, J. (2021). *Information and Experimental Knowledge*. Chicago: The University of Chicago Press, Ltd.

Meyer, G. and Shaheen, S. (2017). *Disrupting Mobility: Impacts of Sharing Economy and Innovative Transportation on Cities*. Berlin: Springer International Publishing.

Miller, G. A. (1956). The Magical Number Seven, Plus or Minus Two: Some Limits on our Capacity for Processing Information. *Psychological Review*, 63, 81–97.

Muller, S. J. (2007). *Asymmetry: The Foundation of Information*. Berlin: Springer-Verlag Berlin Heidelberg.

Negroponte, N. (1995). *Being-Digital*. London: Hodder & Stoughton.

Pintar, J. and Hopping, D. (2023). *Information Science: The Basics*. London and New York: Routledge.

Poster, M. (1990). *The Mode of Information: Poststructuralism and Social Context*. Maiden, MA: Polity Press.

Rajaee, F. (2000). *Globalization on Trial, the Human Condition and the Information Civilization*. Ottawa: Kumarian Press.

Richard, J. (2004). Illustrating the Concept of Quantum Information. *IBM Journal of Research and Development*, 4(1), 79–85.

Roszak, T. (1994). *The Cult of Information: The Folklore of Computers and the True Art of Thinking*. Berkeley: University of California Press.

Rovelli, C. (1996). Relational Quantum Mechanics. *International Journal of Theoretical Physics*, 35(8), 1637–1678.

Schneider, H. (2017). *Creative Destruction and the Sharing Economy, Uber as Disruptive Innovation*. Cheltenham: Edward Elgar Publishing Limited.

Schuster, A. J. ed. (2017). *Understanding Information, From the Big Bang to Big Data*. Cham: Springer International Publishing AG.

Siegfried, T. (2000). *The Bit and the Pendulum: From Quantum Computing to M Theory— The New Physics of Information*. New York: Wiley and Sons.

Shannon, C. E. (1948). A Mathematical Theory of Communication. *Bell System Technical Journal*, 27(7), 379–423.

Shannon, C. E., Sloane, N. J. A. and Wyner, A. D. (1993). *Collected Papers of Claude E. Shannon*. New York: IEEE Press.

Siegfried, T. (2000). *The Bit and the Pendulum: From Quantum Computing to M Theory— The New Physics of Information*. New York: Wiley and Sons.

Sokolov, A. B. (1991). Is Information A Phenomenon, A Function, or an Illusion? Translated by Bai Su. *World Philosophy*, 26(2), 1–8.

Stamper, R. K., (1985). Information: Mystical Fluid or a Subject for Scientific Enquiry? *The Computer Journal*, 28(3), 195–199.

Stephany, A. (2015). *The Business of Sharing*. New York: Palgrave Macmillan.

Stonier, T. (1990). *Information and the Internal Structure of the Universe: An Exploration into Information Physics*. London: Springer-Verlag.

Stratonovich, R. L. (2020). *Theory of Information and Its Value*. Cham: Springer Nature Switzerland AG.

Sumei, C. 成素梅. (2018). The Connotation and Value of Information Civilization in the Age of Big Data 信息文明的内涵及其时代价值. *Academic Monthly* 学术月刊, 62(5), 36–44.

Tianen, W. 王天恩. (2015). Lun Xinxiwenmin信息文明论, *South China Quarterly* 南国学术, 5(3), 93–99.

Timpson, C. G. (2013). *Quantum Information Theory and The Foundations of Quantum Mechanics*. Oxford: Oxford University Press.

———. (2013). *Quantum Information Theory and the Foundations of Quantum Mechanics*. Oxford: Oxford University Press.

Tribes M. and McIrvine, E. C. (1971). Energy and Information. *Scientific American*, 225(3), 179–190.

Toffler, A. (1991). *The Third Wave*. New York: Bantam Books.

Tzafestas, S. G. (2018). *Energy, Information, Feedback, Adaptation, and Self-organization*. Cham: Springer International Publishing AG.

Watson, O. ed. (1976). *Longman Modern English Dictionary*. London: Longman Group Limited.

Wiener, N. (1985). *Cybernetic, or Control and Communication in the Animal and the Machine*. Second edition. Cambridge: The MIT. Press.

———. (1989). *Cybernetics and Society*. Boston: Houghton Mifflin Company.

Wheeler, J. (1990). Information, Physics, Quantum: The Search for Links. In: H. Z. Wojciech ed., *Complexity, Entropy and the Physics of Information*. London and New York: CRC Press. Taylor & Francis Group.

Wierzbicki, A. P. (2006). Technology and Change: The Role of Information Technology in Knowledge Civilization. *Journal of Telecommunications and Information Technology*, 4, 3–14.

Xiao-Gang, W. (2018). Four Revolutions in Physics and the Second Quantum Revolution, a Unication of Force and Matter by Quantum Information. *International Journal of Modern Physics B*, 32(26), 1–24.

Yixin, Z. 钟义信. (2013). *Xinxi Kexue Yuanl* 信息科学原理. Beijing: Beijing University of Posts and Telecommunications.

Zuboff, S. (2014). Obama, Merkel, and the Bridge to An Information Civilization. *Frankfurter Allgemeine Zeitung GmbH*, 17(01), 1–5.

———. (2015). Big Other: Surveillance Capitalism and the Prospects of an Information Civilization. *Journal of Information Technology*, 30, 75–89.

Index

abstraction level 7, 9
academic separatism 2
agent: artificial intelligent 35, 96, 148;
    human 84, 100; information 73, 76–77,
    135, 137, 139–43, 148, 152, 160, 164;
    level of 84; machine 100; receiver-
    source-integrated 21, 36
agricultural civilization 150–51, 157,
    164–65
AI sensors 45
analog encoding 87, 104, 109, 112
animals 37, 45, 76, 93, 132
annihilated 43
anthropological characteristics of
    knowledge 67
anthropology features 49
application scope of information 3
artificial general intelligence: to achieving
    2, 91, 95; core mechanism of vi,
    25, 74, 91, 93, 95; development of
    25, 93; evolve into 91; field of 141;
    interconnection between information
    research and 4; see also information
    research; research on 91
artificial informosomes 35
artificial intelligence (AI): big data and 1,
    3, 4, 18, 20, 74, 75, 85, 94, 95, 141, 156;
    core mechanisms of 94; development of
    4, 20, 67, 86, 91, 93–95, 143, 148, 157,
    169; evolution of 91; exploration of 140;
    extension ability of 2; generalization of
    vi, 93; generation of 18; holistic guan
    zhao of 148; research on 96
artificial intelligence development 20;
    (big) data-driven AI 20, 91, 95;
    human knowledge-driven AI 20, 95;
    information-driven AI 20, 91

Barbieri, M. 46–48, 140
Barwise, J. 78

Basic characteristics of information 22, 24,
    130, 132; research on 22, 23; systematic
    derivation of 22; understanding of
    22, 131
Bawden, D. 138
Béranger, J. 109
Beynon-Davies, R. 6, 146
bidirectional cycle 3, 51, 76, 89–90,
    101, 140, 163, 166–70; between big
    data and information understanding
    86, 88, 94–95; between information
    understanding and the development of
    artificial intelligence 94; mechanism 51;
    multi-level 24; understanding 3
bidirectional interpretation 131
big data: as a collection of digital
    encoding of information 86; context
    of 90; development of 18, 20, 25, 41,
    87–91, 94–95, 150; emergence of
    information in 95; foundation of 91,
    95, 150; as informational existence 86;
    understanding of 26, 89, 96; unfolding
    of information by 26, 156; unveiling of
    information 86, 89–90; value of 143
biological encoding of information 36, 103
biological genes 75, 96, 101–2
biological intelligence 80, 96
from a biological receptivity base to a
    psychological receptivity base 36
bit understanding of information 83
Bohr, N. 55–58
Burgin, M. 77

carbon-based receiver 36
Castells, M. 149
ceiling of information understanding 20,
    25, 48, 123, 151
celestial bodies 43
Chalmers, D. J. 140
Changlin, L. 40

characteristic of receiving and responding to stimuli 45
characteristics of information encoding 22
Charles, P. 98
ChatGPT 2, 20, 91, 95
chemical evolution 43, 48
chemical interaction 43, 44, 133
chemical paradigm 46–47
chemical signals 43
Chinese philosophy 3
coalescence 2
concept series 5, 9–12; collection of concepts with varying extensions 10; of holistic guan zhao 5, 8, 10; *see also* holistic guan zhao（整体观照）; information 9
conceptual encoding 105–8, 110, 112, 114–15; development of 106; nature of 115; Pre- 105–6
conceptual frameworks 11
conceptual issues 8
conceptual speculation 62
conceptual system 21, 65, 83, 84, 107, 110, 114, 115; corresponding 112; distinct from 10; establishing 21, 107; establishment of 115; information conceptual encoding and 107; knowledge as a 110; related to the 112; research on 112; role of 107
condensate physics 132, 136–37
consciousness: "easy problems" of 140; "hard problems" of 140; human consciousness 4; mystery of vii, 2, 4, 45, 74, 91, 170; self- 76, 84
Copernican revolution 73
Copernican transformation 74
correlation 38, 44, 86, 142, 143
co-share 23–24, 144–48, 152–55, 158–59, 165, 169; civilization of 157–59, 169; Foundation of 159
counterfactual communication 33, 43
counting fingers 87
Created ex nihilo of information 22, 49, 126, 132–36, 146–48; characteristic of 134, 148; implications of 135; proof of 133
create something out of nothing 39
creative construction of information 22, 169
Cult of Information 12

Dan, Z. 116
data revolution 90
definition spectrum 9; information 9

descriptive nature of quantum superposition 62
dichotomous approach 6
dichotomous distinction 7
dichotomous framework 76, 78, 80–82; subject and object 65, 76, 78–80
dichotomy abstraction 7, 69
dichotomy concept 7
dichotomy of subject and object 53, 80
Dictionary of Information Science and Technology 99
differentiation in the description and interpretation of quantum theory 66
differentiation of receivers and sources 23, 102
digital age 88, 96
digital data 20, 85, 87, 88, 90, 96, 116
digital encoding 48–49, 83, 86–87, 94, 100, 104, 109–12, 114–16, 124, 127–28, 139; development of 86–87, 89; importance of 109; research on 109
digital technology 14, 87
digitization of matter 90
dilemmas in information research 3
disciplinary boundaries 74
di-share 23, 144–48, 152, 153; bikes, cars 144; diachronic 153; efficiency of 145 physical 148; possibilities of 158; synchronic 153
di-share and co-share 23–24, 145, 152, 169
dissociation between the use and understanding 17
DNA 14, 43, 81, 101, 102, 127, 153; carried from 31; controlled by 43; encoded in 46; genes in 36, 81, 100–102, 114, 158
double-slit experiment 54, 119
Dretske, F. 8, 16–17, 29–30, 37, 41, 105, 107, 111–13, 128

Eigen, M. 3, 26
electrical currents 83
electrical pulse encoding 13
electronic receptive system 42
electronic receptor 41
emergence 22, 132–33, 136–41, 147–48; collective 138–39; connotations of 138; informational 22, 136–40, 147; irreducible 132, 137; irreversible 132, 137–38; level of 132, 136; material 22, 136–39, 147; mechanism of 136, 139; nature of information 49; origin of 137;

study of 132, 137, 139; theory 132; types (form) of 22, 138
Emperor's new clothes, the 5, 12, 16
energy encoding 103, 114
enigma of the human brain 50
Entanglement 20, 64, 67, 81, 99, 134
entitative and informational interaction 42
entity interaction 41–42
epistemological consequences 7
epistemological implication 5, 66
era-specific conditions 4
everyday definition 9
everyday use 79
evolution of information, the 3, 21, 76–77, 87, 140, 141, 152
experiential encoding 105–06, 114
experimental arrangements 54, 55, 58, 65, 70, 92
expert system 20, 91
explanation and interpretation 140
explanatory gap 66

family resemblance concept 7, 10
Fen Xiang (分享) 23, 144
Floridi L. 2, 10, 12, 27, 31, 39–40, 87
flow of signal 30, 35
Fresco, N. 99
Fritz Mark Lup' 14
fundamental approaches to information understanding 19

general information science 5
generalizations of receptive relation 107
generalized by dichotomy 7
generating' information 3, 26
genetic information 14, 31
Gleick, J. 103
gong xiang (共享) 23, 144
graphic encoding 106, 114
guan zhao(观照) 3, 7, 92, 116

Hartley, R. 128
Hawking, S. 124
Heisenberg, W. 56
higher-level concepts 3, 7, 11, 105, 108, 113
higher-level wholeness 23
high holistic level 6
Hofkirchner, W. 19
holism 23, 49, 139
holistic guan zhao(观照) 3, 7, 9–12, 14, 21, 60, 143, 157; *see also* holistic perspective; concepts and theoretical

3, 7–8, 11–12, 60, 135; concept series of 5, 8, 10–12; concept system 10; of information 60, 90, 92, 94, 116, 135; of information civilization 149, 156–57, 169–70; level of 3, 11, 90, 148; process of 12; rational 116; theoretical 3, 10; *see also* holistic guan zhao; of theory 59, 101; understanding of abstract concepts 9; understanding of concrete things 11–12, 60, 77
holistic nature of information 4, 8
holistic understanding 4, 18
human-dependent character 37
human entanglement as receivers 20
human existence 49, 76, 86, 93, 134, 143, 160–62, 169
human knowledge 20, 52, 91, 134; -driven AI 20, 91, 95; *see also* artificial intelligence; level of 25, 78; limitations of 64; machine representation of 20; nature of 147; and the nature of theories 108
human life 6, 169
human-like intelligence 20, 25, 52, 64, 78, 91, 95, 108, 134, 147
human sensory characteristics 58, 63, 66, 92
human situation in understanding information 19, 76

ideational encoding 21, 83, 100–01, 104–17, 131, 135; development of 100, 106; forms of 106; Research on 115; role of 115
ideational systems 106, 111
ideational veiling of information 75, 80, 83–84, 86, 96–97, 111
ideationalization 21, 82
ignore information by attributing receptive relation to object properties 71
indirect existence 88
individuality origin from the specific existence of information 6
industrial civilization 149–52, 157, 165, 169
information: abstract noun, an 30, 47, 60, 68; assignment of the receiver 35; concrete existence of 6–7; content of exchanged with the outer world 28; dynamic process, a 40; an effect, a process, or a relation 39; eliminating uncertainty 3, 28–29; entropy 34; field, a 34; form/pattern 33; higher-level

existence 16; metaphor, a 47; negative entropy 34; ontological characteristic of 147; order 34; organization 33; property of matter, a 32, 33; punched cards, patterns on records, and tracks on tapes 85; representation of events 85; resembles flowing water 30; structure 33, 47
    information and information encoding 22, 81, 96, 98, 102; adhesion between 20; confusion between 99, 128; difference between 47, 48; relationship between 83
information biology 4, 117
information carrier 82
information channel 13, 17, 27, 64, 83, 93
information civilization 1, 4, 16, 23–25, 84, 90, 94–95, 148–70; age of 1, 3, 8, 24, 90, 159–67; characteristics of 148, 153, 156; context of 94, 156; co-shared 154; development of 5, 24, 90, 94, 96, 151–57, 141, 148–50, 161–62, 165–69; foundation of 1, 153, 157; holistic guan zhao of 156–57, 169–70; level of 149, 152, 155; nature of 164; research on 150–51
    information concept: cross-disciplinary 17; diversity of 6; and its utilization 17; level of 10, 14; sublevel 9; understanding of 11, 25, 52; unified 6, 11, 14; universal (utility, acceptance) philosophical 7, 10; use of 14; *see also* usage over ownership
information conservation 25, 49, 74, 117–18, 124, 126, 134 dilemma of 3, 25, 74, 130; law of, the 117; misconception of 123; misunderstanding of 118; none-4, 117, 134–35; PIIMO Understanding of 124
information decoding 98, 100–02
information definition 7–9, 14, 35, 39, 109, 111, 128; clear 44; in communication science 27; *see also* information in communication science; everyday 9; fundamental 86; general 86; mathematical 29, 86; multitude of 9, 14; philosophical 7, 34; precise 94, 157; scientific 9; Shannon's 109; unified 3, 7–8, 26; universally (widely) accepted 2, 6, 15; various 1, 26, 80; Wiener's 28, 110–11
information evolution 48, 77, 81, 84, 103; manifestations of 140; process of 138; result of 84; stage of 84

information feedback 76, 170
information flow 30, 31, 34, 39, 87
information grounding 118, 123, 127–29
information in communication science 13, 17, 29; definition of 27, 29; *see also* information definition; theory of 11; understanding of 28, 31
information "Lushan." 41
information mechanism 137, 140–41, 146–47; bidirectional cyclic 21, 111; *see also* bidirectional cycle; irreducibility of 140
information myth 15
information needs 89, 142–43, 155–57, 161; commonality of satisfaction and germination 23–24, 148, 155, 160, 162; concept of 15; development of 161; level of 143, 155–56, 169; nature of 89, 143, 147, 169; research on 26; significance of 89
information obfuscation 20
information objectivism 37
information paradigm shift 46, 74
information philosophy 33, 75
information principle 116, 118; principle of identity for information-matter in operation (piimo), the 21–22, 117, 121; *see also* information mechanism, material coding
information process 21, 25, 35, 69, 76, 103, 114–15, 117, 119, 124–25, 127–29; level of 21, 132; quantifying 128; scope of 125
information production 21, 77, 88, 89, 102, 155, 162
Information Quantum Mechanics 39, 64, 70
information quantum mechanics 39, 64, 70
information receiver 73, 76, 96, 103, 131, 143–44; potential 144
information research 1, 7–10, 15–19, 25–27, 31, 37–38, 76, 90–91, 94–99, 111, 114–15; development of 5, 10, 17, 71, 75, 112–13, 137; dilemmas in 3; framework of 19; holistic guan zhao for 3, 7; at the level of integration of science and philosophy 14, 25, 27, 63; *see also* integration of science and philosophy; principle in 129; quantum 59–60, 68, 134; Winding Path of 18
information revolution 90, 94
information source (sender) 21, 27, 31, 34, 102, 144–45; potential (latent) 21; primary 21; secondary 21; sender 17, 35, 45, 88
information storage 102

information technology: development of 1, 3–5, 16, 19–20, 36, 41, 72, 74–75, 80, 84–90; foundation of 74; level of 94; unfolding of information within 85, 94
information theory 1–3, 11, 13, 18, 28, 30, 34, 60, 63, 120, 170; context of 124; Dretske's 29; Floridi's 10, 87; general 7; philosophical 2, 76; quantum 51, 59–60; scientific 2, 68, 75–76, 79; semantic 29; Shannon's 2, 11, 29, 86; starting point of 40; unified 7, 10, 26, 28
Information Theory: New Trends and Open Problems 40
information unfolding 5, 46, 74, 89, 154, 169
information world 88, 96
informational activities of plants 17, 131
informational existence 20, 80, 86, 88, 130
Informational interpretation of quantum physics 60
informational nature of quantum phenomena 19, 51, 59
Informational Root of Quantum Paradoxes 60
informational way 7, 74, 80, 96, 156, 157, 161, 162, 169, 170
informational ways 7, 80
information-physical meaning of quantum mechanics 59
informatization of matter 90, 101, 102, 153
integrated disciplinary nature 9
integrated research 25
integration of science and philosophy vi, 1, 3, 4, 8, 12, 14, 16, 26, 27, 59, 86
intellectual tension 3
intelligent civilization 4
intentionality 77, 78
interaction of matter/energy 41
interactions in receptivity 41
interaction with chemical properties 42
interdisciplinary integration 11
Internet of Things (IoT) 158

Janich, P. 5, 15

Kelly, K. 37, 41, 139, 142
Klous, S. 150, 161
knotting records 87
Knowledge and the Flow of Information 30

Landauer, R. 32, 84, 124–26
language games 7
large language models (LLM) 2, 20, 91, 95, 145
Laura, F. 64

Lombardi, O. 28, 60, 69
Longo, G. 15, 40
lower-level concepts 11

machine receive 35, 36
machine representation of human knowledge 20
manifestation of information 73, 83, 141; material 81, 138; quantum phenomena 20, 51, 55, 58, 72, 74, 80, 96; *see also* quantum phenomena
material civilization 24, 148, 151–53, 156–57, 159, 163–65, 168–69; conditions of 167; context of 157; development of 152; forms of 169; foundation of 151, 153, 156
material encoding 21–22, 48, 82–84, 87, 100–04, 107, 114–15; convert into 115; imitations of 107; research on 21
material encoding and ideational encoding 100, 109, 115, 130
material needs 23, 89, 143, 153, 161, 162, 169
material operations of information 22
material resources 22, 24, 89, 134, 135, 152–55, 158, 159, 161, 163–65
material veiling of information 73, 75, 80–84, 88, 96
materialization 21, 82, 90, 100–104, 114, 115, 161, 169; of data 90; of information 90, 101–3, 114, 161
mathematical description 26, 62, 66
Mathematical formalization 86
Mathematical Theory of Communication 2, 13, 28
matter physics 67, 120, 132
McIrvine, E. C. 108
meaning-enduing mechanism 21
memory 100–02, 105–06, 112–13, 115, 125; biological 100, 112, 114; characteristic of 114; development of 102; human 101, 113
message 13–14, 16–17, 27, 31, 35; composed of 31; an information understanding of 28; a series of signals 28
molecular biology 46
Muller, S. J. 17, 86
mutation 138–39, 147
mystery of life, the vii, 2, 4, 45, 74, 91, 170
Mysterious Existence 5, 12
mystery of quantum signal 29

natural receptivity 36
Negroponte, N. 151

new reality 1
new-quality interactions 41
noise 3, 26, 27
non-existence 21, 118, 147

object encoding 103–04
object quantification methods for
   information quantification 29
object-coding 87
objective understanding of information
   34, 36
observation: effect vii, 55, 65, 70,
   92; informational nature of 61; an
   informational process of establishing
   receptive relation 53; medium of 70, 71;
   theory-ladenness of 113
ontological implications 22
ontological premise 11
ontological stipulation 11
ontology of biological information 46
operational identity 49, 122
operational significance 22, 122
operationalism 122–23
organismic receptive interaction 101
organismic receptivity 19, 45, 81, 101,
   103, 133
organismic receptivity and sensory
   receptivity 19, 23, 45, 49, 93, 134, 144

paradigm gap 76
paradigm shift 1, 5, 18, 19, 40, 52, 73, 76,
   79; achieving 148; basic 39; challenging
   4; corresponding 130; the deepest 20;
   effects of 94; fundamental paradigm
   shift vi, vii, 1, 19, 26, 76, 141; of
   information 74, 19, 170; involvement
   of 133; problem of 132; profound 48;
   relation 40
paradigmatic dilemma 84
paradox 25, 31–32, 35, 49, 55, 76,
   118, 126, 134; of atomic physics
   57; development 5; in information
   understanding 117;of paradigm 84;
   properties 146; quantum 60–62, 71; of
   wave-particle duality 55
perceptual abilities 20
phenomenon-level problem 5
pheromone 30, 33, 41, 43, 44, 75, 96
philosophical information theory 2, 76
philosophical reflections on information
   1, 87
philosophical understanding of information
   2–4, 29
physical carriers (mediums) of signals 69

physical characteristic understanding 22
physical interactions and signal
   interactions 43
physical imprint 102
physical reactions, biological receptivity,
   and psychological receptivity 35
physical sources of information 81
physical understanding of information 21,
   22, 32, 70, 119, 122, 124, 125
physical world 77, 79–81, 110, 115, 126
plants 17, 19, 43, 45, 47, 76, 102, 131
point of Archimedes 44
Poster, M. 150
potential secondary source 85, 102
potential signal 43
potential source 21, 30, 82, 85, 89
potential sources (latent sources) 21
potential sources of information 82
Preskill, J. 125
primarily operates at the level of utilizing
   information 17
Primary information sources 21
probabilistic nature of signal processing 30
process of differentiating objectivity and
   subjectivity 37
properties of matter and energy 78
pure objectivity 59
purely matter interaction and receptive
   interaction 41

Qbit ocean 81
qualitative understanding of information as
   relation 39
Quantifiability 2
quantifiable nature 26
quantification of information 30, 39,
   118, 128
quantum Cheshire Cat 61–63, 71
quantum communication 33, 43, 68
quantum information 29–30, 33, 51–53,
   55, 60, 64, 67–71, 119, 126, 134;
   characteristic of 59; Fog around 67;
   meaningful theory of 71; mystery of
   29; nature of 29; non-matter property
   of 30; Qubit as 33; research on 52, 65,
   68; theory of 71; understanding of 52,
   55, 70
Quantum Manifestation of Information
   51, 96
quantum observation 53–55, 61–64, 69–70,
   92, 119; barrier of receptivity 54;
   information in 55, 69, 71; Information
   Stripping 53; medium in 119; object in
   54, 58; separation of information 55

quantum paradoxes 60, 61, 71
quantum phenomena vii, 19–20, 45, 51–71, 73–74, 92–93, 95, 119, 134; definition of 58; informational nature of 19, 51, 59–60; information and 20, 65, 68, 81, 92; manifestation of information 20, 51, 55, 58, 72, 74, 80, 83, 96, 138; relation nature of 61, 71; understanding of 67
quantum physics 4, 19, 46, 51–52, 55–71, 77, 123; *see also* information as receptive relation; and classical physics 65–67, 71, 92, 118, 130, 134; development of 74; informational approach to 51, 64; information and 51–52; *see also* information theory; presentation of information 19, 51, 58, 64, 71, 77, 92, 95, 100; research on 63; status of 51; understanding of 19, 52, 58, 64, 70

Rajaee, F. 150
receiver-assigning understanding of information 34–38, 40, 68, 92, 122, 131–32; cognitive origin of the 36; importance of the 34, 37
receptive activity 134
receptive interaction: condensation of 104, 114; effects in quantum observation 71; feature of 43; irreversibility of 140; organismic 101; process of 45, 85, 134; property of 42; and quantum phenomena 19, 42
receptive process 43
receptive relation: abstraction of 114; achieving 48; based on matter/energy 45, 82; context of 101, 137, 146; created ex nihilo of 132, 136; *see also* created ex nihilo; development of 44, 91, 115; dynamic process of 49; effects of 102; establish 43, 45, 48, 53–54, 62, 65, 69, 71, 77, 82, 85, 101, 107, 110, 113, 116, 119; formation of 137; foundation of 106; generalizations of 107; green color as 46; holistic guan zhao of 92; idealization of 104, 115; inherent characteristic of 134, 137; level of 67, 93; manifestation of 141; materialization of 101–02; meaning of 48; nature of information 22, 60, 71, 89, 102, 136, 141, 142, 144, 146, 147; nature of quantum phenomena 19, 51, 59–60, 71; observation effect as 63, 92; organismic receptive relation 93, 114, 135, 141; perspective of 77, 125; process relation or relation process 45; quantum phenomena as 20, 58, 61,

65–66; rebuilding of 98, 100, 123, 145; Research on 49; role of 21; sensory receptive relation 141; understanding of 60, 77, 116, 135, 138, 147; unique type of relation, a 45; various 106
receptivity barrier 54
reciprocity of information 22–24, 49, 133, 141–44, 146–48, 153, 155; anthropological 24, 170; concept of 141; exploration of 141; foundation of 142; of holistic guan zhao 60; nature of 142; significance of 141; strengthening of 159; superimposition of 24, 43, 170; understanding of 142–43; unfolding of 161
reductionism 11, 23, 49, 139–40; reductionist physics 137
relation paradigm transition 49
relational existence, a 86, 89, 133
relational items 83
relational ontology 49
Relational Quantum Mechanics 39
relational understanding of information 39–40
relationality 47, 82–83
relationship between matter and information 73, 119
relationship with holism and reductionism 23, 139
Resultant information obfuscation 20
Robinson, L. 138
Roszak T. 12–14

Schrödinger's Cat 61–62, 71
Schuster, A. J. 4, 6–8, 28
science of communication, the 2
scientific and philosophical contexts 86
scientific community, the 1
scientific understanding of information, the 2–3, 18, 26, 29
secondary information source 21, 84, 101, 104, 116, 119, 124, 130, 145; potential 85, 102
self-evident 91, 118, 155
self-involvement 18, 38
sensory receptivity 19, 45, 77, 108, 110; development of 103; interactions of 133
shadowless lamp 12
Shannon, C. E. 2
shareability 22, 23, 147; co- 146; of information 22, 144–45, 148; of matter and energy 152
shift from the material paradigm to the information paradigm 19, 94–95
signal carriers 69–71, 82

signal-carrying understanding of information 27–31, 37, 39, 88; basis of 31; characteristic of the 30
signal processing 30, 86
silicon-based receiver 36
solipsism tendency of the receiver-assigning understanding 38
source-based information generation formula 28
source-preexisting understanding of information 31–34, 37, 39–40, 55, 81, 89, 92, 109, 134; challenge of 33
specialized receptive organ 36
speculative theoretical systems 12
Stamper, R. 146
strong-current interaction 41
subatomic scale as a signal 42
subjectivity of information 36
subject-object relations in the realm of information 36
Sumei, C. 150
symbolic encoding 105, 108–09, 112, 114–15
systems groups 139

theoretical perspective connected to ontology 11
theory of relativity vi, 2, 19, 65, 73
thought stipulation 113; of "noumenon" 11
Timpson, C. G. 33, 60, 68, 92
Toffler, A. 150
Total Baseball: The Ultimate Baseball Encyclopedia 125
Tribes, M. 108
Tzafestas, S. G. 6

ultimate truth of abstract universality 7
unavoidable disturbance 53
unveiling of information 32, 73–74, 85, 88–96, 112, 127; artificial intelligence 91; data 86–90, 95–96; information technology 92, 96; material and ideational 92
usage over ownership 155

Veiling and Unveiling of Information, The 73
veiling of information 20, 52, 73–75, 80, 85, 90; ideational 83–84, 86; level of 92; material 47, 80–84, 88, 96; material and ideational dual 80, 95; by material paradigm 20, 124; *see also* human entanglement as receivers; by matter and energy 73;
Venus flytrap 45

Watson, O. 99
weak-current interaction 41
Wheeler, J. 110
Wielaard, N. 150, 161
Wiener, N. 14, 15, 26, 28, 32, 34, 44, 103, 110, 151, 157
Wittgenstein, L. 7
Wolf, M. J. 99
Wu, K. 88

Xiaogang, W. 81

Yixin, Z. 2, 9, 28, 109

Zubof, S. 150

For Product Safety Concerns and Information please contact our EU
representative GPSR@taylorandfrancis.com
Taylor & Francis Verlag GmbH, Kaufingerstraße 24, 80331 München, Germany

www.ingramcontent.com/pod-product-compliance
Lightning Source LLC
Chambersburg PA
CBHW060306220326
41598CB00027B/4248

9 781032 778051